ADVANCED NANOTUBE AND NANOFIBER MATERIALS

NANOTECHNOLOGY SCIENCE AND TECHNOLOGY

Additional books in this series can be found on Nova's website
under the Series tab.

Additional e-books in this series can be found on Nova's website
under the e-books tab.

MATERIALS SCIENCE AND TECHNOLOGIES

Additional books in this series can be found on Nova's website
under the Series tab.

Additional e-books in this series can be found on Nova's website
under the e-books tab.

NANOTECHNOLOGY SCIENCE AND TECHNOLOGY

ADVANCED NANOTUBE AND NANOFIBER MATERIALS

A. K. HAGHI

AND

G. E. ZAIKOV

EDITORS

Nova Science Publishers, Inc.

New York

For permission to use material from this book please contact us:
Telephone 631-231-7269; Fax 631-231-8175
Web Site: http://www.novapublishers.com

NOTICE TO THE READER

Library of Congress Cataloging-in-Publication Data

Advanced nanotube and nanofiber materials / editors, A.K. Haghi, G.E. Zaikov.
p. cm.
Includes index.
ISBN 978-1-62081-170-2 (soft cover)
1. Nanotubes. 2. Nanostructured materials. I. Haghi, A. K. II. Zaikov, G. E. (Gennadii Efremovich), 1935-
TA418.9.N35A32875 2012
620.1'15--dc23
2012005444

Published by Nova Science Publishers, Inc. † New York

CONTENTS

PREFACE

Nowadays, the promising field of nanotechnology has a revolutionary impact on science and technology. Although the development of nanotechnology occurred in the late eighties, the idea of nanotechnology was introduced in 1959, when Feynman, in his talk on the possibility to precisely manipulate atoms and molecules commented, *"But I am not afraid to consider the final question as to whether, ultimately in the great future we can arrange the atoms the way we want; the very atoms, all the way down!"*

Thereafter, the field of nanotechnology was created by Eric Drexler by expanding Feynman's vision of molecular manufacturing with contemporary developments in understanding protein function. Drexler discussed the possibility of molecular manufacturing as a process of fabricating objects with specific atomic specifications using designed protein molecules. Although the term "nanotechnology" is used by Taniguchi in 1974, in a different context, Drexler is credited as being the first person to use the word nanotechnology in his famous book *Engines of Creation -The Coming Era of Nanotechnology*.

Although the terms *nanomaterial* and *nanocomposite* represent new and exciting fields in materials science, such materials have actually been used for centuries and have always existed in nature. However, it is only recently that the means to characterize and control structure at the nanoscale have stimulated rational investigation and exploitation. A nanocomposite is defined as a composite material where at least one of the dimensions of one of its constituents is on the nano-metre size scale. The term usually also implies the combination of two (or more) distinct materials, such as a ceramic and a polymer, rather than spontaneously phase-segregated structures. The challenge and interest in developing nanocomposites is to find ways to create macroscopic components that benefit from the unique physical and mechanical properties of very small objects within them. Natural materials such as bone, tooth, and nacre are very good examples of the successful implementation of this concept, offering excellent mechanical properties compared to those of their constituent materials. Such composites actually exhibit beautifully organized levels of hierarchical structure from macroscopic to microscopic length scales and provide a powerful motivation for improving our processing control. Currently, we are striving to understand the behavior of just the smallest building blocks in such materials, which are the natural versions of nanocomposites. Significantly, two contrasting phases are often combined: a hard nanoscale reinforcement (such as hydroxyapatite or calcium carbonate) is embedded in a soft, usually protein-based, matrix. Although the composite character of these materials itself plays a crucial role, the question remains as to why the nano-metre scale is so important.

The term Nano, a factor of 10^{-9}, has its origin in the Greek word nanos, meaning dwarf. A nanostructure is an object of size between molecular and microscopic structures. It is a product at the molecular scale. However, nanoparticles are very tiny aggregations of atoms; they are bigger than most of the molecules.

Generally, there are two processes to create nanoscale materials from atoms and molecules. **First is the "bottom-up" process that creates nanoscale** materials from atoms and molecules. The second process is the "top-down" process that creates nanoscale materials from their macro-scale counterparts.

Nanostructured materials are used in several applications like catalysis, electronics, separation technologies, sensors, information storage, drug delivery systems, diagnostics, energy batteries, fuel cells, solar cells, etc. The prospective of nanomaterials in biomedical and industrial applications for human health and environment are now well established. Moreover, the nanoclusters, nanoparticles, nanotubes, nanoporous materials, nanowires, hybrid nanocomposites, etc., are used in every branch of science and technology. Nanoscience is its interdisciplinary nature—its practice requires researchers to cross the traditional boundaries between the experimental and theoretical fields of chemistry and physics, materials science and engineering, biology and medicine, to work together. Various research fields including physics, chemists, material scientists, and engineers are involved in this research.

Nanochemistry, the first step in nanotechnology, is a new branch of nanoscience that permits controlling chemical parameters in order to grow nano-objects. Thus, it attracts tremendous attention in recent researches.

Scientists for the first time applied the principles of chemistry to the bottom-up synthesis of nanomaterials.

The aim of a synthetic nanochemist is to design nanoscale building blocks of desired shape, size, composition and surface structure. At the nanolevel, the so-called quantum effects can be significant, fascinating and potentially scientifically very rewarding innovative ways of carrying out chemical reactions are possible.

The recent advances in the field of nanochemistry are discussed in several literatures. The dependence of chemico-physical properties on the size of the nanoparticles are also studied based on thermodynamics, electrochemistry, optical spectra and magnetic properties. In this review, we will focus on the applications of nanocompounds in various fields of science and technology.

The concept of creating both structural and functional multi-phase nanocomposites with improved performance is currently under development in a wide variety of metallic, ceramic, and polymeric matrices, although the emphasis to date has been on polymeric systems.

Similarly, the filler particles can be organic or inorganic with a wide range of material compositions and structures. The resulting composites generally exhibit a number of enhanced properties, so that the material cannot easily be classified as a structural or functional composite. The term *reinforcement*, as opposed to plain *filler*, is equally frequently used for the nanoscale component, without a clear distinction.

CARBON NANOTUBES (CNTs)

Carbon nanotubes (CNTs) have attracted particular interest because they are predicted, and indeed observed, to have remarkable mechanical and other physical properties. The combination of these properties with very low densities suggests that CNTs are ideal candidates for high-performance polymer composites; in a sense, they may be the next generation of carbon fibers. Although tens or hundreds of kilograms of carbon nanotubes are currently produced per day, the development of high-strength and high-stiffness polymer composites based on these carbon nanostructures has been hampered so far by the lack of availability of high-quality (high crystallinity) nanotubes in large quantities. In addition, a number of fundamental challenges arise from the small size of these fillers. Although significant advances have been made in recent years to overcome difficulties with the manufacture of polymer nanocomposites, processing remains a key challenge in fully utilizing the properties of the nanoscale reinforcement. A primary difficulty is achieving a good dispersion of the nanoscale filler in a composite, independent of filler shape and aspect ratio. Without proper dispersion, filler aggregates tend to act as defect sites, which limit the mechanical performance; such agglomerates also adversely influence physical composite properties such as optical transmissivity.

A variety of synthesis methods now exist to produce carbon nanotubes and nanofibers. However, these carbon nanostructures differ greatly with regard to their diameter, aspect ratio, crystallinity, crystalline orientation, purity, entanglement, surface chemistry, and straightness. These structural variations dramatically affect intrinsic properties, processing, and behavior in composite systems. However, it is not yet clear which type of nanotube material is most suitable for composite applications, nor is there much theoretical basis for rational design. Ultimately, the selection will depend on the matrix material, processing technology, and the property enhancement required. Thus, in order to interpret the data obtained for nanotube composites, and to develop the required understanding, it is essential to appreciate the range of nanotube materials available.

NANOMEDICINE

Nanomedicine has been an important part of nanotechnology from the very beginning. Nanochemistry works with materials at the atomic level and has many potential applications for medical science. The application of nanotechnology to medicine concerns the use of nanomaterials to develop novel therapeutic and diagnostic drugs known as "nanomedicines." Nanoscientists have developed nanodrugs to reach specific molecular targets on diseased cells and have been used in various experimental and clinical conditions. The medical application involves diagnostic and therapeutic applications, and a large deal of this research concerns malignant disease. Various approaches have been tried to effectively reach the cancer cell.

Nanomaterials, having some unique chemico-physical properties, such as ultra-small size, large surface area-to-mass ratio, and high reactivity, can be used to overcome some of the limitations of the traditional therapeutic and diagnostic agents.

The nanomedicines are in the similar size-range as viruses, DNA and proteins. Used with various well-chosen molecules, recent medicinal nanochemistry decorated the surface are of the nanomedicine so that the immune system can't recognize them. Thus, the nanodrugs can easily reach their target more efficiently. The nanoparticles are also designed to overcome the blood brain barrier and dermal tight junctions. Nanoparticles are synthesized to carry drugs and to release them at a site of disease.

Nanoparticles are designed nowadays not only for the specific delivery but also for the penetration in solid tumors. The synthesized nanoparticles can penetrate the lesion due to the leaky constitution of neovasculature in malignant tumors.

Nanoparticles consist of an inorganic core of superparamagnetic materials coated with polymer and are used as contrast agents in magnetic resonance imaging for diagnostic applications and therapy monitoring.

It is widely known that the gold nanoparticles find wide scientific use and applications.

Because of some unique features, mesoporous silica-based nanostructured platform holds great promise for medicinal chemistry. The platforms have the potential to form ordered pore network for finely controlling the drug load and release kinetics. Moreover, having a high pore volume, they can incorporate high dosages of drugs inside the nanochannels and having a high surface area, they can manipulate the conjugation of the drugs with different therapeutic and biotargeting molecules.

It is also demonstrated that these platforms prevent the drug from its enzymatic degradation prior to reaching the target site.

The potential applications of the multi-functional nanocomposite nanoparticles for simultaneous fluorescence, magnetic resonance imaging (MRI), pH-sensitive drug release, etc., were also fabricated.

Thus, Nanotechnology can provide the technical power and tools that will enable those developing new diagnostics, therapeutics, and preventives to keep pace with today's explosion in knowledge. With nanomedicine, we might be able to stop a disease like cancer even before it develops.

With such technology, nanomedicine has the potential to increase the lifespan of human beings, and hence the nanotechnology will radically change the way we diagnose, treat and prevent diseases.

OIL INDUSTRY

Nowadays, the fundamental concepts of nanochemistry are applied for the synthesis of a wide variety of useful chemicals such as pharmaceuticals, commercial polymers or catalysts having potential impact in the oil industry [26].

NANOELECTRONICS

Nanotechnology is one of the most active research areas that encompass a number of disciplines such as electronics, biomechanics and coatings including civil engineering and construction materials.

The small size of nanoparticles gives these particles "unusual" structural and optical properties with applications in catalysis, electro-optical devices etc. Having remarkable electronic properties and many other unique characteristics, carbon nanotubes attract both experimentalists and theorists to study the properties of these materials. The usefulness of the single-walled carbon nanotubes (SWNETs) in nanoscale electronics and lightweight materials are now well established.

Because of their chemico-physical importance, the nanoparticles of coinage metals are now integral part of the nanotechnology.

Currently, researchers have developed a nanoparticle ink composed of a water-based solution mixed with a high concentration of silver nanoparticles that could make flexible printed electronics. This printing method greatly increases the complexity possible and limiting breakage and other manufacturing problems.

To avoid the difficulty of depositing the printing materials in the molten form, scientists introduced a system in which an ink-jet print head deposits a nanoparticle colloid ink to print three-dimensional (3-D) metallic structures. As the process of the ink-jet printing is noncontact, the nanomaterials are added to the bulk layer-by-layer to form 3-D structures.

Current high-technology production processes based on nanotechnology have developed a carbon nanotube-based crossbar memory called Nano-RAM.

The quantum dots can be used for the construction of lasers, which are cheaper than the traditional semiconductor laser. The other benefit of quantum dot laser is that it offers a higher beam quality than conventional laser diodes.

The effect magnetoresistance can be significantly amplified for nanosized objects. The Giant Magneto-Resistance effect has led to a strong increase in the data storage density of hard disks and made the gigabyte range possible.

Although quantum computing is still in its infancy, scientists are now engaged in preparing novel quantum computers, which enable the use of fast quantum algorithms. The Quantum computer with quantum bit memory space (Qubit) is capable of doing several computations at the same time.

ENVIRONMENTAL MONITORING

Nanotechnology also has the potential to rectify the effects causing the environmental pollution by creating materials and products that will not only directly advance our ability to detect, monitor, and clean-up environmental contaminants but also help us avoid the environmental pollution.

It is demonstrated that nanomaterials such as silica-titania nanocomposites are useful to remove elemental mercury from vapors. It is also demonstrated by several scientists that nanostructured silica can sorb other metals, such as lead and cadmium generated in combustion environments.

The maturation of nanotechnology has revealed it to be a unique and distinct discipline rather than a specialization within a larger field. Its textbook cannot afford to be a chemistry, physics, or engineering text focused on nano. It must be an integrated, multi-disciplinary, and specifically nanotextbook. This book surveys the field's broad landscape, exploring the physical basics such as nanorheology, nanofluidics, and nanomechanics as well as industrial concerns such as manufacturing, reliability, and safety. The authors then explore the vast range of nanomaterials and systematically outline devices and applications in various industrial sectors.

The book then presents the tools of nanotechnology that can build, image, and manipulate nanostructures to build materials and devices.

This book bridges the gap between detailed technical publications that are beyond the grasp of non-specialists and popular science books, which may be more science fiction than fact. It provides a fascinating, scientifically sound treatment, accessible to engineers and scientists outside the field and even to students at the undergraduate level. The book concludes with a look at some cutting-edge applications and prophecies for the future.

This book builds a solid background in characterization and fabrication methods, while integrating the physics, chemistry, and biology facets. The book focuses on applications and examining engineering aspects.

A. K. Haghi
University of Guilan, Iran
G. E. Zaikov
Russian Academy of Sciences

In: Advanced Nanotube and Nanofiber Materials   ISBN: 978-1-62081-170-2
Editors: A. K. Haghi and G. E. Zaikov   © 2012 Nova Science Publishers, Inc.

Chapter 1

CARBON NANOTUBES

A. K. Haghi[*]

University of Guilan, Iran

1.1. INTRODUCTION

In 1991, Japanese researchers studied sediment formed at the cathode during the spray of graphite in an electric arc. Their attention was attracted by the unusual structure of the sediment consisting of microscopic fibers and filaments. Measurements made with an electron microscope showed that the diameter of these filaments do not exceed a few nanometers and a length of from one to several microns.

Having managed to cut a thin tube along the longitudinal axis, the researchers found that it consists of one or more layers, each of which represents a hexagonal grid of graphite, which is based on hexagon with vertices located at the corners of the carbon atoms. In all cases, the distance between the layers is equal to 0.34 nm, which is the same as that between the layers in crystalline graphite.

Typically, the upper ends of tubes are closed by multi-layer hemispherical caps; each layer is composed of hexagons and pentagons, reminiscent of the structure of half a fullerene molecule.

The extended structure consisting of rolled hexagonal grids with carbon atoms at the nodes are called nanotubes.

---

[*] Haghi@Guilan.ac.ir.

Lattice structure of diamond and graphite are shown in Figure 1.1. Graphite crystals are built of planes parallel to each other, in which carbon atoms are arranged at the corners of regular hexagons. The distance between adjacent carbon atoms (each side of the hexagon), between adjacent planes 0.335 nm.

Each intermediate plane is shifted somewhat toward the neighboring planes, as shown in the Figure 1.1.

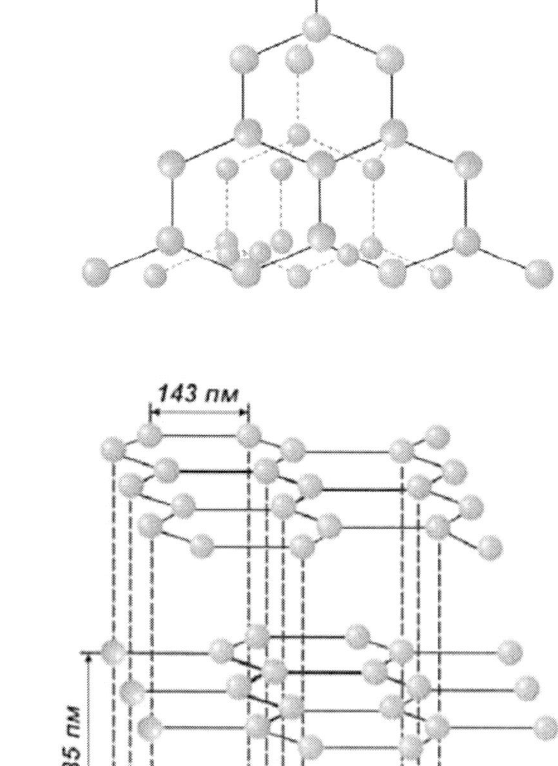

a)

b)

Figure 1.1. The structure of the diamond lattice a) and graphite b).

The elementary cell of the diamond crystal is a tetrahedron in the center and four vertices of which are carbon atoms. Atoms located at the vertices of a tetrahedron form a new center of a tetrahedron, and thus, are also surrounded by four atoms each, etc. All the carbon atoms in the crystal lattice are located at equal distance (0.154 nm) from each other.

Nanotubes rolled into a cylinder (hollow tube) graphite plane, which is lined with regular hexagons with carbon atoms at the vertices of a diameter of several nanometers (Figure 1.2). Nanotubes can consist of one layer of atoms (single-wall nanotubes- SWNT) and represent a number of "nested" into one another layer pipes (multi-walled nanotubes – MWNT).

Nanostructures can be collected not only from individual atoms or single molecules but the molecular blocks. Such blocks or elements to create nanostructures are graphene, carbon nanotubes and fullerenes.

1.2. GRAPHENE

Graphene is a single flat sheet, consisting of carbon atoms linked together and forming a grid; each cell is like a bee's honeycombs (Figure 1.2). The distance between adjacent carbon atoms in graphene is about 0.14 nm.

Graphite, from which are made slates of usual pencils, is a pile of graphene sheets (Figure 1.3). Graphenes in graphite is very poorly connected and can slide relative to each other. So, if you conduct the graphite on paper, then after separating graphene from sheet, the graphite remains on paper. This explains why graphite can write.

Figure 1.2. Schematic illustration of the graphene.

Figure 1.3. schematic illustrations of the three sheets of graphene.

1.3. CARBON NANOTUBES

Many perspective directions in nanotechnology are associated with carbon nanotubes.

Carbon nanotubes: a carcass structure or a giant molecule consisting only from carbon atoms.

Carbon nanotube is easy to imagine, if we imagine that you fold up one of the molecular layers of graphite-graphene (Figure 1.5).

Figure 1.4. Carbon nanotubes.

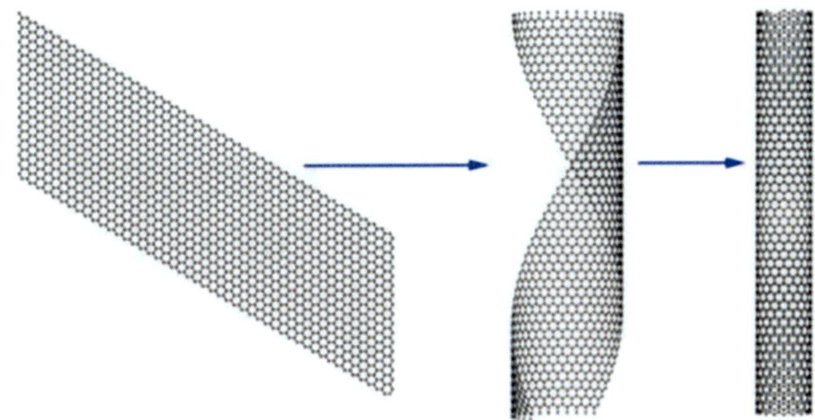

Figure 1.5. Imaginary making nanotube (right) from the molecular layer of graphite (left).

Nanotubes formed themselves, for example, on the surface of carbon electrodes during arc discharge between them. At discharge, the carbon atoms evaporate from the surface and connected with each other to form nanotubes of all kinds: single, multi-layered and with different angles of twist (Figure 1.6).

The diameter of nanotubes is usually about 1 nm, and their length is a thousand times more, amounting to about 40 microns. They grow on the cathode in perpendicular direction to surface of the butt. Occurring so is called self-assembly of carbon nanotubes from carbon atoms. Depending on the angle of folding, the nanotube can have a high as that of metals, conductivity, and can have properties of semiconductors.

Carbon nanotubes are stronger than graphite, although made of the same carbon atoms, because the carbon atoms in graphite are in the sheets. And everyone knows that folding into a tube sheet of paper is much more difficult to bend and break than a regular sheet. That's why carbon nanotubes are strong. Nanotubes can be used as very strong microscopic rods and filaments, as Young's modulus of single-walled nanotube reaches values of the order of 1-5 TPa, which is much more than steel! Therefore, the thread made of nanotubes the thickness of a human hair is capable of holding down hundreds of kilos of cargo.

It is true that at present, the maximum length of nanotubes is usually about a hundred microns—which is certainly too small for everyday use. However, the length of the nanotubes obtained in the laboratory is gradually increasing—now scientists have come close to the millimeter border. So there is every reason to hope that in the near future, scientists will learn how to grow a nanotube length in centimeters and even meters!

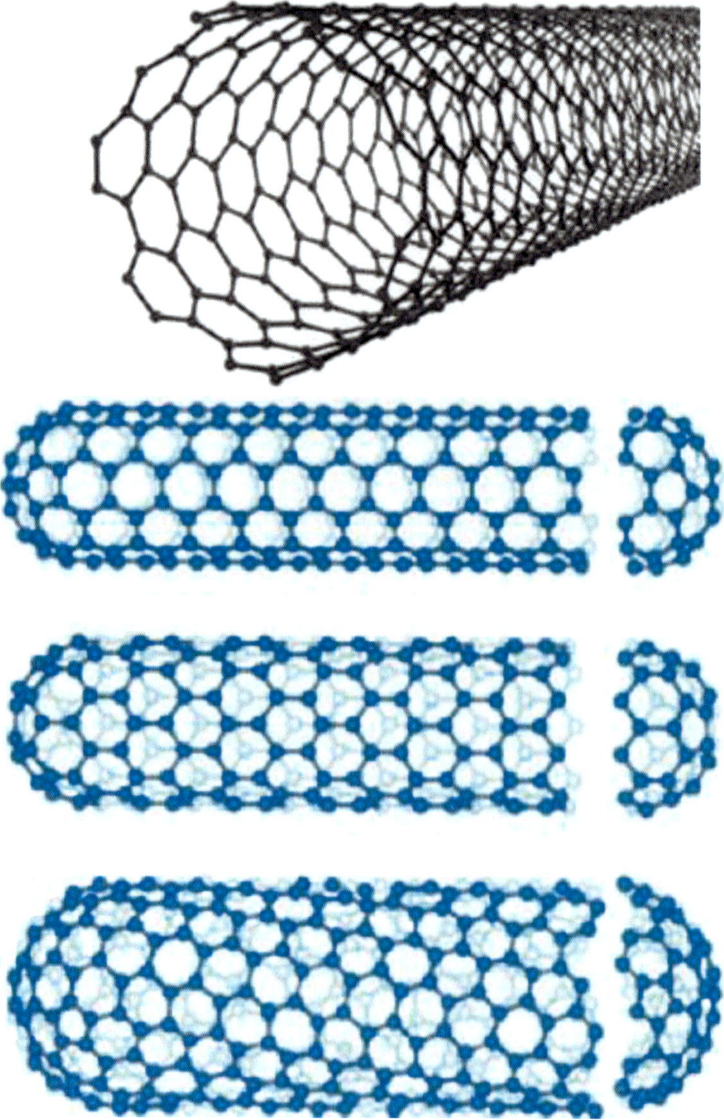

Figure 1.6. Schematic representation of a single-layer carbon nanotubes, on the right (top to bottom)—two-ply, straight and spiral nanotubes.

1.4. FULLERENES

The carbon atoms, evaporated from a heated graphite surface, connecting with each other, can form not only of the nanotube, but also other molecules, which are closed convex polyhedra, for example, in the form of a sphere or ellipsoid. In these molecules, the carbon atoms located at the vertices of regular hexagons and pentagons that make up the surface of a sphere or ellipsoid.

All of these molecular compounds of carbon atoms called fullerenes on behalf of the American engineer, designer and architect, R. Buckminster Fuller, whose domes were used for construction of its buildings, pentagons and hexagons (Figure 1.7), which are the main structural elements of the molecular carcasses of all of fullerenes.

The molecules of the symmetrical and the most studied fullerene consisting of 60 carbon atoms (C_{60}), form a polyhedron consisting of 20 hexagons and 12 pentagons and resemble a soccer ball (Figure 1.8). The diameter of the fullerene is about 1 nm.

Figure 1.7. Biosphere of Fuller (Montreal, Canada).

Figure 1.8. Schematic representation of the fullerene C_{60}.

1.5. CLASSIFICATION OF NANOTUBES

The main classification of nanotubes is conducted by the number of constituent layers.

Single-walled nanotubes: the simplest form of nanotubes. Most of them have a diameter of about 1 nm in length, which can be many thousands of times more. The structure of the nanotubes can be represented as a "wrap" a hexagonal network of graphite (graphene), which is based on hexagon with vertices located at the corners of the carbon atoms in a seamless cylinder. The upper ends of the tubes are closed by hemispherical caps; each layer is composed of six pentagons, reminiscent of the structure of half of a fullerene molecule. The distance d between adjacent carbon atoms in the nanotube is approximately equal to nm.

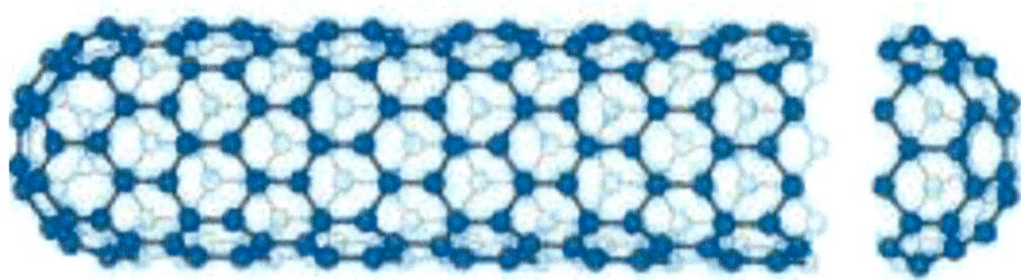

Figure 1.9. Graphical representation of single-walled nanotube.

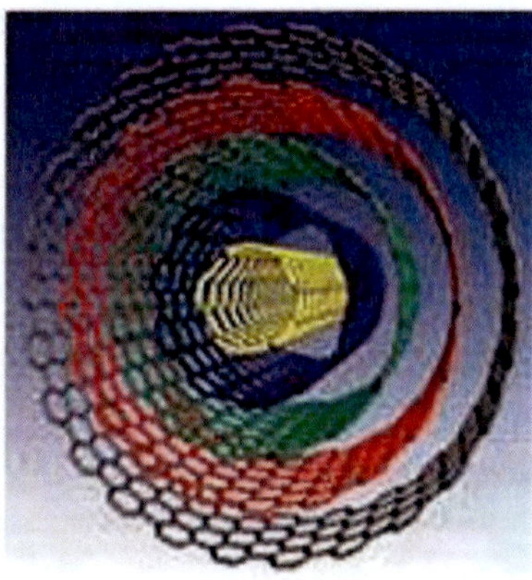

Figure 1.10. Graphic representation of a multi-walled nanotube.

Multi-walled nanotubes consist of several layers of graphene stacked in the shape of the tube. The distance between the layers is equal to 0.34 nm, which is the same as that between the layers in crystalline graphite.

Due to its unique properties (high fastness (63 GPa), superconductivity, capillary, optical, magnetic properties, etc.), carbon nanotubes could find applications in numerous areas:

- Additives in polymers;
- Catalysts (autoelectronic emission for cathode ray lighting elements, planar panel of displays, gas discharge tubes in telecom networks);
- Absorption and screening electromagnetic waves;
- Transformation of energy;
- Anodes in lithium batteries;
- Keeping of hydrogen;
- Composites (filler or coating);
- Nanosondes;
- Sensors;
- Strengthening of composites;
- Supercapacitors.

For more than a decade, carbon nanotubes, despite their impressive performance characteristics, have been used, in most cases, for scientific research. These materials are not yet able to gain a foothold in the market, mainly because of problems with their large-scale production and un-competitive prices.

To date, the most developed production of nanotubes has been in Asia, with a production capacity that is two to three times higher than in North America and Europe combined. Japan dominates in the production of MWNT. Manufacturing North America mainly focused on the SWNT. In the coming years, China will surpass the level of production of the U.S. and Japan, and by now, a major supplier of all types of nanotubes, according to experts, could be South Korea.

6. CHIRALITY

Chirality is a set of two integer positive indices (n, m), which determines how the folds the graphite plane and how many elementary cells of graphite at the same time fold to obtain the nanotube.

From the value of parameters are distinguished

- direct (achiral) high-symmetry carbon nanotubes
    - armchair
    - zigzag or
- helical (chiral) nanotube

Figure 1.11a shows a schematic image of the atomic structure of graphite plane—grapheme—and shows how from it can be obtained the nanotube. The nanotube is folded up with the vector connecting two atoms on a graphite sheet. The cylinder is obtained by folding this sheet so that the beginning and end of the vector were combined. That is, to obtain a carbon nanotube from a graphene sheet, it should turn so that the lattice vector has a circumference of the nanotube in Figure 1.11b. This vector can be expressed in terms of the basis vectors of the elementary cell graphene sheet $\vec{R} = n\vec{r_1} + m\vec{r_2}$. Vector $\overline{R}$, which is often referred to simply by a pair of indices (n, m), called the chiral vector. It is assumed that $n > m$. Each pair of numbers represents the possible structure of the nanotube.

In other words, the chirality of the nanotubes indicates the coordinates of the hexagon, which as a result of folding the plane has to coincide with a hexagon, located at the beginning of coordinates (Figure 1.12).

Many of the properties of nanotubes (for example, zonal structure or space group of symmetry) strongly depend on the value of the chiral vector.

Chirality indicates what property a nanotube has—a semiconductor or metallicheskm. For example, a nanotube (10, 10) in the elementary cell contains 40 atoms and is the type of metal, whereas the nanotube (10, 9) has already in 1084 and is a semiconductor (Figure 1.13).

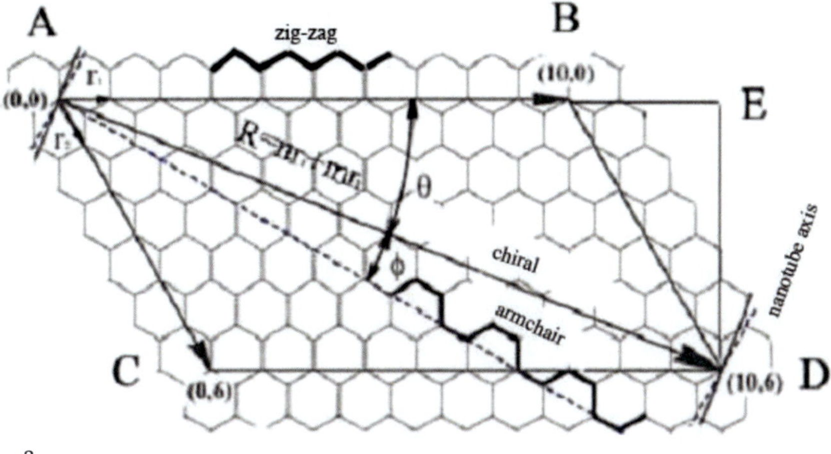

a.

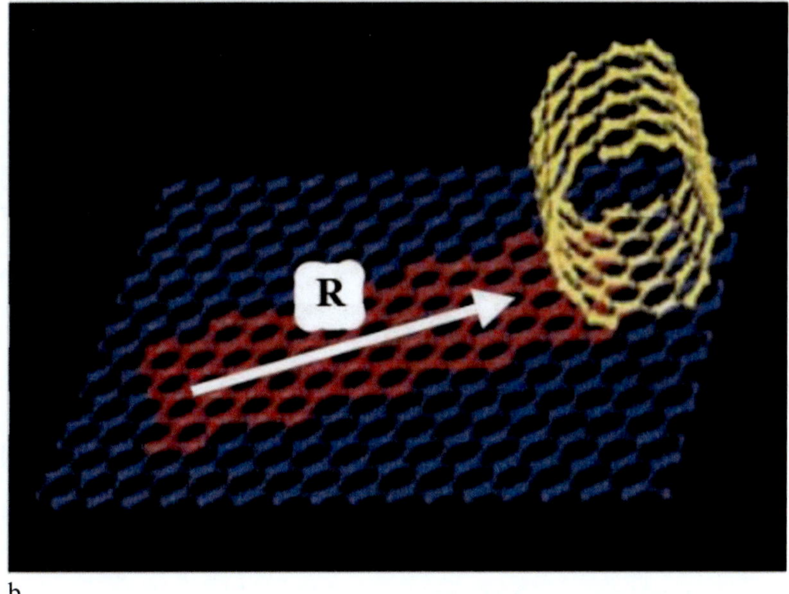

b.

Figure 1.11. Atomic structure of graphite plane.

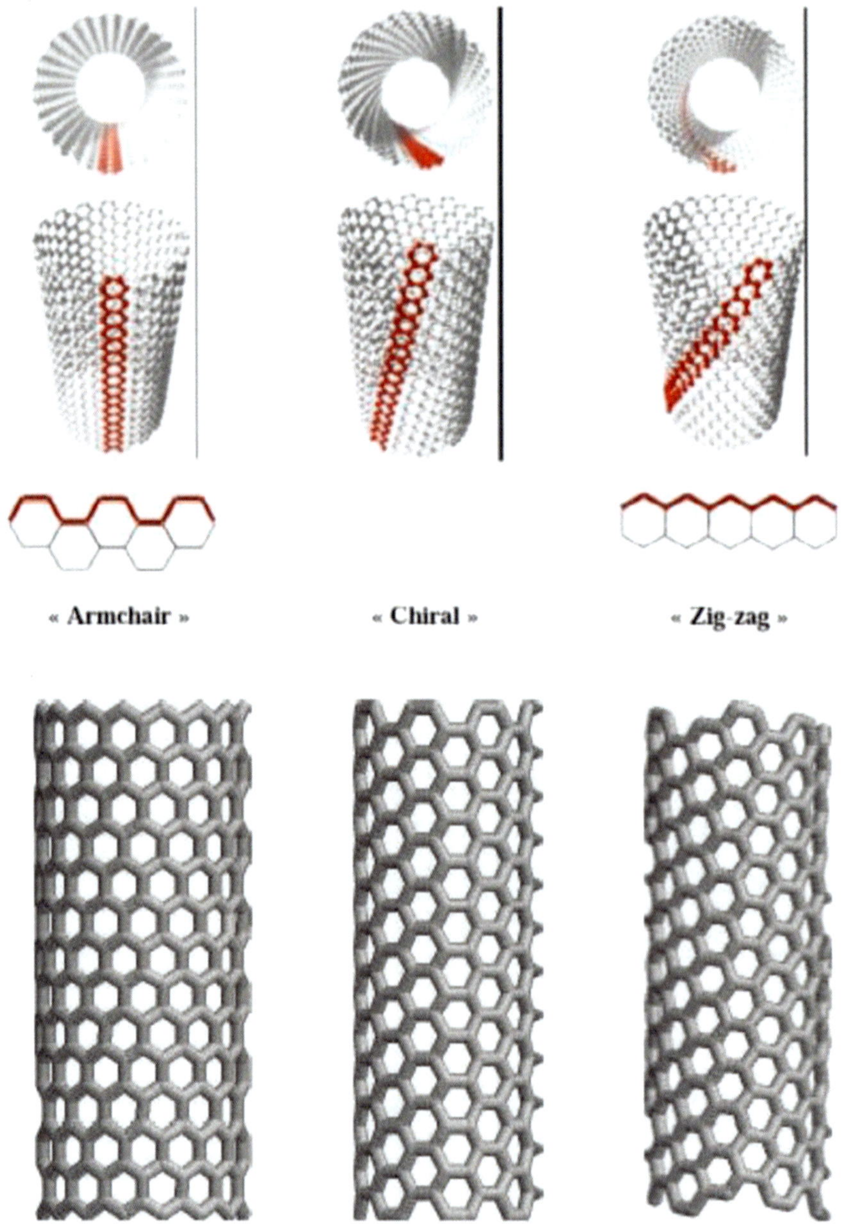

Figure 1.12. Single-walled carbon nanotubes in different chirality, Left to right: the zigzag (16, 0), armchair (8, 8) and chiral (10, 6) carbon nanotubes.

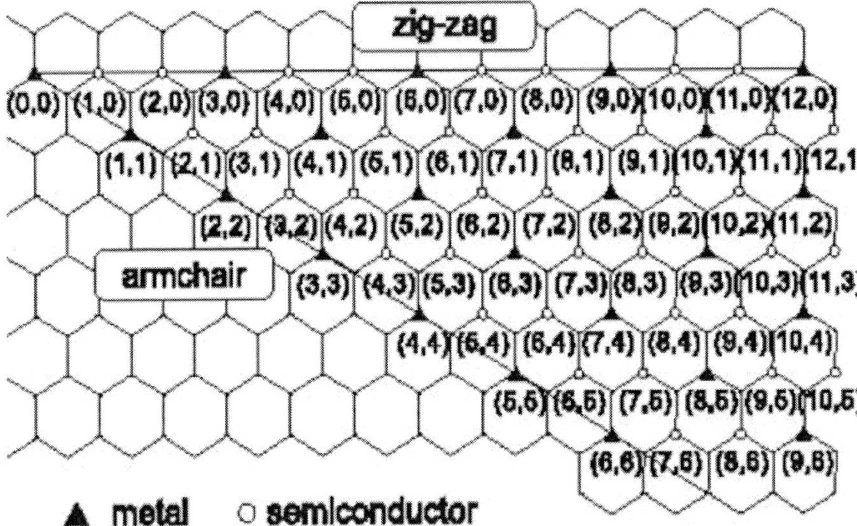

Figure 1.13. The scheme of indices of lattice vector tubes having semiconductor and metallic properties.

If the difference is divisible by three, then these CNTs have metallic properties. Semimetals are all achiral tubes such as "chair." In other cases, the CNTs show semiconducting properties. Just type chair CNTs are strictly metal.

1.7. Diameter, Chirality Angle and the Mass of Single-walled Nanotube

Indices of single-walled nanotube chirality unambiguously determine its diameter. Therefore, the nanotubes are typically characterized by a diameter and chirality angle. Chiral angle of nanotubes is the angle between the axis of the tube and the most densely packed rows of atoms. From geometrical considerations, it is easy to deduce relations for the chiral angle and diameter of the nanotube. The angle between the basis vectors of the elementary cell (Figure 1.14) is equal to 60^0.

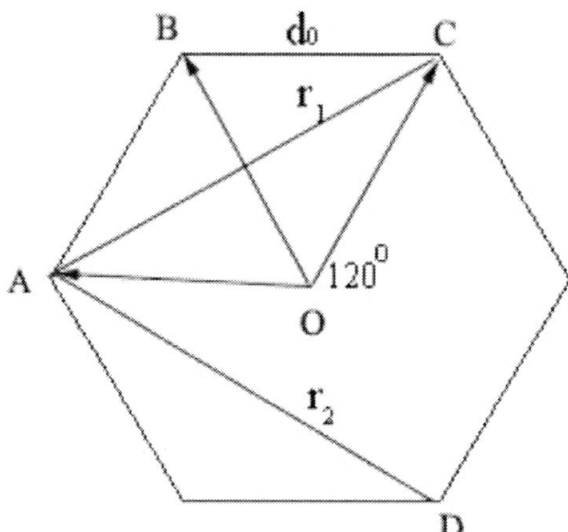

Figure 1.14. The elementary cell.

As we know from trigonometry, $AC^2 = OA^2 + OC^2 - 2OA \cdot OC \cdot \cos 120^0$. As $OA = OC = d_0$, a $r_1 = r_2 = AC$, we have

$$r_1 = r_2 = \sqrt{3} \cdot d_0, \tag{1.1}$$

where - distance between neighboring carbon atoms in the graphite plane. Thus, the basis vectors и of the elementary cell of graphene are $\left|\vec{r_1}\right| = \left|\vec{r_2}\right| = 0{,}244 нм$.

Now consider the parallelogram in Figure 1.11a.
According to (1.1), we have

$$AB = CD = \sqrt{3}d_0 n, \quad AC = BD = \sqrt{3}d_0 m \tag{1.2}$$

Angle $\angle CAB = 60^0$, and $\angle ABD = 120^0$, therefore
$R^2 = 3n^2 d_0^2 + 3m^2 d_0^2 - 2 \cdot 3mn d_0^2 \cos 120^0$, from which we obtain
$$R = \sqrt{3}d_0 \sqrt{n^2 + m^2 + mn}$$

Taking into account that $R = \pi \cdot d$, then to determine the diameter of the nanotube, we obtain the expression

$$d = \frac{\left|\vec{R}\right|}{\pi} = \sqrt{3\left(m^2 + n^2 + mn\right)} \cdot \frac{d_0}{\pi} \qquad (1.3)$$

When we have

$$d = \frac{3nd_0}{\pi}$$

Below in Table 1.1 the values of the diameters of nanotubes of different chirality are shown.

Thus, knowing the chirality can be found and possible relations and n (Table 1.2). The minimum diameter of the tube is close to 0.4 nm, which corresponds to the chirality (3, 3), (5, 0), (4, 2). Unfortunately, the objects of that the diameter of the least stable. Of single-walled nanotube, was one most stable with chirality indices (10, 10); its diameter is equal 1.35 nm.

We derive a formula for determining the mass of the nanotube with diameter d, length L.

Table 1.1. diameters of nanotubes of different chirality

(n, m)	d, nm	(n, m)	d, nm
(3,2)	0,334	(10,8)	1,232
(4,2)	0,417	(10,9)	1,298
(4,3)	0,480	(11,3)	1,007
(5,0)	0,394	(11,6)	1,177
(5,1)	0,439	(11,10)	1,434
(5,3)	0,552	(12,8)	1,375
(6,1)	0,517	(14,13)	1,844
(7,3)	0,701	(20,19)	2,663
(9,2)	0,801	(21,19)	2,732
(9,8)	1,161	(40,38)	5,326

Table 1.2. CNT with of different chirality

CNT (n,m)	Diameter CNT, nm	Chirality
(4,0)	0,33	zigzag
(5,0)	0,39	
(6,0)	0,47	
(7,0)	0,55	
(8,0)	0,63	
(9,0)	0,70	
(10,0)	0,78	
(11,0)	0,86	
(12,0)	0,93	
(3,3)	0,40	armchair
(4,4)	0,56	
(5,5)	0,69	
(6,6)	0,81	
(7,7)	0,96	
(8,8)	1,10	
(4,1)	0,39	chiral
(4,2)	0,43	
(7,1)	0,57	
(6,3)	0,62	
(9,1)	0,75	
(10,1)	0,82	
(6,7)	0,90	

The area of the elementary area—a parallelogram with vertices at the centers of four neighboring hexagons (Figure 1.15) with base and height is equal $S_{n\pi} = \dfrac{3\sqrt{3}}{2} d_0^2$.

The total area of the nanotube is πdL. Consequently, the number of elementary areas is equal $\pi dL / S_{n\pi}$. At the same time, in each elementary site contains two carbon atoms. Consequently, the number of carbon atoms in the tube is twice more than the number of elementary areas that can fit on the surface. Therefore, the mass of a carbon nanotube is equal to:

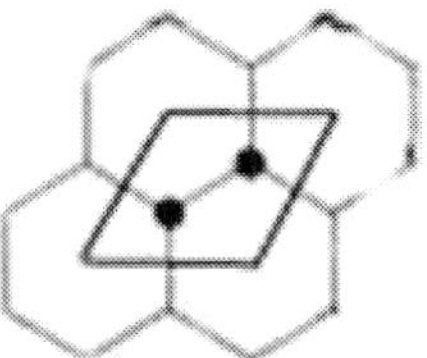

Figure 1.15. The elementary area of graphene.

$$m_T = 2m_C \frac{\pi L d}{S_{n\pi}} = \frac{4\sqrt{3}\pi \cdot dL}{9d_0^2} m_C, \qquad (1.4)$$

where $m_C = 12$—mass of carbon atoms.

To determine the chiral angle from a right triangle, we obtain

$$\sin\theta = \frac{DE}{R}, \quad \cos\theta = \frac{AE}{R} = \frac{\sqrt{3}nd_0 + BE}{R}$$

If we take into consideration that $\angle EDB = 30^0$, we see that $BE = \frac{\sqrt{3}}{2}md_0$, consequently,

$$\sin\theta = \frac{3md_0}{2R}, \quad \cos\theta = \frac{\sqrt{3}d_0(n + m/2)}{R}$$

From these equalities, we obtain the relation between the chiral indices and angle θ:

$$\theta = arctg\left(\frac{\sqrt{3}m}{2n + m}\right) \qquad (1.5)$$

When we have

$$\theta = arctg\frac{\sqrt{3}}{3}$$

1.8. OUTLOOKS

Although the terms *nanomaterial* and *nanocomposite* represent new and exciting fields in materials science, such materials have actually been used for centuries and have always existed in nature. However, it is only recently that the means to characterize and control structure at the nanoscale have stimulated rational investigation and exploitation. A nanocomposite is defined as a composite material where at least one of the dimensions of one of its constituents is on the nanometre-size scale. The term usually also implies the combination of two (or more) distinct materials, such as a ceramic and a polymer, rather than spontaneously phase-segregated structures. The challenge and interest in developing nanocomposites is to find ways to create macroscopic components that benefit from the unique physical and mechanical properties of very small objects within them. Natural materials such as bone, tooth, and nacre are very good examples of the successful implementation of this concept, offering excellent mechanical properties compared to those of their constituent materials. Such composites actually exhibit beautifully organized levels of hierarchical structure from macroscopic to microscopic length scales and provide a powerful motivation for improving our processing control.

Currently, we are striving to understand the behaviour of just the smallest building blocks in such materials, which are the natural versions of nanocomposites. Significantly, two contrasting phases are often combined: a hard nanoscale reinforcement (such as hydroxyapatite or calcium carbonate) is embedded in a soft, usually protein-based, matrix.

Although the composite character of these materials itself plays a crucial role, the question remains as to why the nanometre scale is so important. From a simple mechanical point of view, the situation in such biocomposites is quite familiar: the matrix transfers the load via shear to the nanoscale reinforcement. A large length-to-diameter (aspect) ratio of the mineral reinforcement compensates for the low modulus of the soft protein matrix, leading to an optimised stiffness of the composite. The fracture toughness of such biocomposites, on the other hand, hinges on the ultimate tensile strength of the reinforcement. Crucially, the use of a nanomaterial allows access to the maximum theoretical strength of the material, since mechanical properties become increasingly insensitive to flaws at the nanoscale. This observation is an extension of the classic approach to strong materials, namely to reduce the dimensions until critical flaws are excluded. At the nanoscale, highly crystalline reinforcements are used in which all but the smallest atomistic defects can be eliminated. It is clear that a high aspect ratio must be maintained in order to ensure suitable stress transfer. This general concept of exploiting the inherent properties of nanoscaled materials is not limited to the mechanical properties of a material, since a wide range of physical properties also depend on defect concentrations. In addition, the small size scale can generate inherently novel effects through, for example, quantum confinement or through the dramatic increase in interfacial area. The concept of creating both structural and functional multi-phase nanocomposites with improved performance is currently under development in a wide variety of metallic, ceramic, and polymeric matrices, although the emphasis to date has been on polymeric systems. Similarly, the filler particles can be organic or inorganic with a wide range of material compositions and structures.

The resulting composites generally exhibit a number of enhanced properties, so that the material cannot easily be classified as a structural or functional composite. The term *reinforcement*, as opposed to plain *filler*, is equally frequently used for the nanoscale component, without a clear distinction. Carbon nanotubes (CNTs) have attracted particular interest because they are predicted, and indeed observed, to have remarkable mechanical and other physical properties. The combination of these properties with very low densities suggests that CNTs are ideal candidates for high-performance polymer composites; in a sense, they may be the next generation of carbon fibres. Although tens or hundreds of kilograms of carbon nanotubes are currently produced per day, the development of high-strength and high-stiffness polymer composites based on these carbon nanostructures has been hampered so far by the lack of availability of high-quality (high-crystallinity) nanotubes in large quantities. In addition, a number of fundamental challenges arise from the small size of these fillers. Although significant advances have been made in recent years to overcome difficulties with the manufacture of polymer nanocomposites, processing remains a key challenge in fully utilizing the properties of the nanoscale reinforcement. A primary difficulty is achieving a good dispersion of the nanoscale filler in a composite, independent of filler shape and aspect ratio. Without proper dispersion, filler aggregates tend to act as defect sites, which limit the mechanical performance; such agglomerates also adversely influence physical composite properties such as optical transmissivity. When dispersing small particles in a low viscosity medium, diffusion processes and particle-particle and particle-matrix inter-actions play an increasingly important role as the diameter drops below 1 μm. It is not only the absolute size but rather the specific surface area of the filler, and the resulting interfacial volumes, which significantly influence the dispersion process. These regions can have distinctly different properties from the bulk polymer and can represent a substantial volume fraction of the matrix for nanoparticles with surface areas of the order of hundreds of m2/g. The actual interphase volume depends on the dispersion and distribution of the filler particles, as well as their surface area. In traditional fibre composites, the interfacial region is defined as the volume in which the properties deviate from those of the bulk matrix or filler.

REFERENCES

[1] M. Ziabari, V. Mottaghitalab, S. T. McGovern and A. K. Haghi, *Chim. Phys. Lett.*, 25, 3071 (2008).

[2] M. Ziabari, V. Mottaghitalab, S. T. McGovern and A. K. Haghi, *Nanoscale Research Letter*, 2, 297(2007).

[3] M. Ziabari, V. Mottaghitalab and A. K. Haghi, *Korean J. Chem. Eng.*, 25, 919 (2008).

[4] M. Ziabari, V. Mottaghitalab and A. K. Haghi, *Korean J. Chem. Eng.*, 25, 923 (2008).

[5] M. Ziabari, V. Mottaghitalab and A. K. Haghi, *Korean J. Chem. Eng.*, 25, 905 (2008).

[6] A. K. Haghi and M. Akbari, *Physica Status Solidi*, 204, 1830 (2007).

[7] M. Kanafchian, M. Valizadeh and A.K. Haghi, *Korean J. Chem. Eng.*, 28, 428 (2011).

[8] M. Kanafchian, M. Valizadeh and A.K. Haghi, *Korean J. Chem. Eng.*, 28, 763 (2011).

[9] M. Kanafchian, M. Valizadeh and A.K. Haghi, *Korean J. Chem. Eng.*, 28, 751 (2011).

[10] M. Kanafchian, M. Valizadeh and A.K. Haghi, *Korean J. Chem. Eng.*, 28, 445(2011).

[11] A. Afzali, V. Mottaghitalab, M. Motlagh, A.K. Haghi, *Korean J. Chem. Eng.*, 27, 1145(2010).

[12] Z. Moridi, V. Mottaghitalab, A.K. Haghi, *Korean J. Chem. Eng.*, 28, 445(2011).

[13] A.K. Haghi, *Cellulose Chem. Technol.*, 44, 343 (2010)

[14] Z. Moridi, V. Mottaghitalab, A.K. Haghi, *Cellulose Chem. Technol.*, 45, 549 (2011)

In: Advanced Nanotube and Nanofiber Materials   ISBN: 978-1-62081-170-2
Editors: A. K. Haghi and G. E. Zaikov   © 2012 Nova Science Publishers, Inc.

Chapter 2

RECENT PROGRESS ON CARBON NANOTUBE/NANOFIBER COMPOSITES

A. K. Haghi[*]
University of Guilan, Iran

1. INTRODUCTION

Recently, the words "nanobiocomposites" or "biopolymer nanocomposites" is most frequently observed in environmentally friendly research studies. The synthetic polymers have been widely used in various applications of nanocomposites. However, they become a major source of waste after use due to their poor biodegradability. On the other hand, most of the synthetic polymers are without biocompatibility in *vivo* and *vitro* environments. Hence, scientists were interested in biopolymers as biodegradable materials [1]. Later, several groups of natural biopolymers such as polysaccharide, proteins, and nucleic acids were used in various applications [2]. Nevertheless, the use of these materials has been limited due to relatively poor mechanical properties. Therefore, researcher efforts have been made to improve the properties of biopolymers as a matrix by using of reinforcement [3].

Chitosan (CS) is a polysaccharide biopolymer that has been widely used as a matrix in nanobiocomposites. Chitosan represents high biocompatibility

---

[*] Haghi@Guilan.ac.ir.

and biodegradibility properties, although these biopolymers have an essential requirement to additional material with high mechanical properties [4]. Following discovery of carbon nanotube, results of characterization represented unique electrical and mechanical properties. Thereby, many research studies have focused on improving the physical properties of biopolymer nanocomposites by using the fundamental behavior of carbon nanotubes [5].

It is the aim of this review to summarize recent advances in the production of carbon nanotubes/chitosan nanocomposites by several methods. Specifically, we will discuss our recent work in preparing CNTs/CS nanofiber composites by using of electrospinning method.

2. BIOPOLYMERS

Biomaterial has been defined as biocompatibility materials with the living systems. The biocompatibility implies the chemical, physical (surface morphology), and biological suitability of an implant surface to the host tissues. S. Ramakrishna et al. reviewed various biomatcrials and their applications over the last 30 years. They represented applications of biopolymers and their biocomposites in medical applications [6]. These materials can classify to natural and synthetic biopolymers. Synthetic biopolymers have been provided cheaper with high mechanical properties. The low biocompatibility of synthetic biopolymers compared with natural biopolymers such as polysaccharides, lipids, and proteins have led to great attention being paid to the natural biopolymers. On the other hand, the natural biopolymers usually have weak mechanical properties. Therefore, many efforts have been done for improving their properties by blending some filler [7].

Among the natural biopolymers, polysaccharides seem to be the most promising materials in various biomedical fields. These biopolymers have various resources, including animal origin, plant origin, algal origin, and microbial origin. Among various polysaccharides, chitosan is the most usual due to its chemical structure [8].

2.1. Chitosan

Chitin (Figure 1.) is the second-most abundant natural polymer in the world and extracted from various plants and animals [9]. However, derivations of chitin have been noticed because insolubility of chitin in aqueous media. Chitosan (Figure 2.) is deacetylated derivation of chitin with the form of free amine. Unlike chitin, chitosan is soluble in diluted acids and organic acids. Polysaccharides are containing 2-acetamido-2-deoxy-β-D-glucose and 2-amino-2-deoxy-β-D-glucose. Deacetylation of chitin converts acetamide groups to amino groups [10]. Deacetylation of degree (DD) is one of the important effective parameters in chitosan properties and has been defined as "the mole fraction of deacetylated units in the polymer chain" [11].

Chitosan could be suitably modified to impart desired properties due to the presence of the amino groups. Hence, a wide variety of applications for chitosan have been reported over the recent decades. Table 1. shows chiotosan applications in a variety of fields and their principal characteristics. The high biocompatibility [12] and biodegradability [13] of chitosan yield most potential applications in biomedical [14].

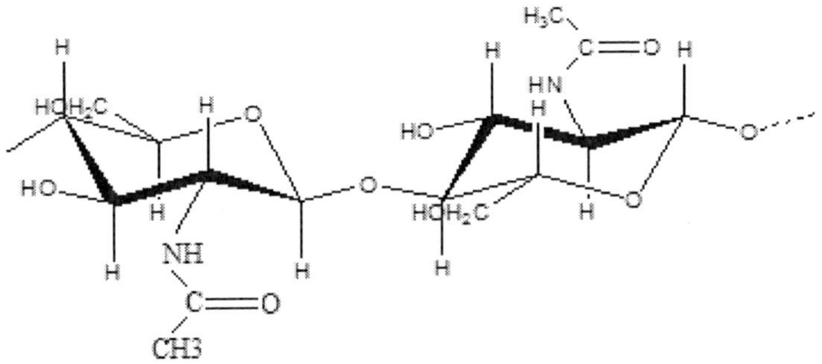

Figure 1. Structure of chitin.

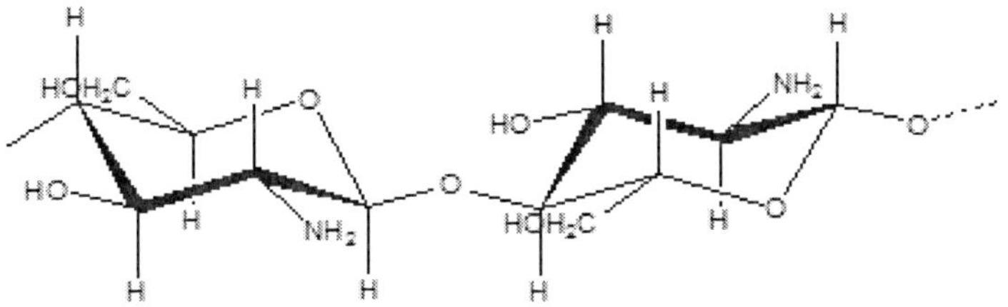

Figure 2. Structure of chitosan.

Table 1. chiotosan applications in variety fields and their principal characteristics

Chitosan application		Principal characteristics	Ref
water engineering		metal ionic adsorption	[15]
biomedical application	biosensors and immobilization of enzymes and cells	biocompatibility, biodegradability to harmless products, nontoxicity, antibacterial properties, gel-forming properties and hydrophilicity, remarkable affinity to proteins	[16]
	antimicrobial and wound dressing	wound-healing properties	[17]
	tissue engineering	biocompatibility, biodegradable, and antimicrobial properties	[18]
	drug and gene delivery	biodegradable, nontoxicity, biocompatibility, high charge density, mucoadhesion	[19]
	orthopedic/perio dontal application	antibacterial	[20]
Photography		resistance to abrasion, optical characteristics, film-forming ability	[21]
cosmetic application		fungicidal and fungi static properties	[22]
food preservative		biodegradability, biocompatibility, antimicrobial activity, non-toxicity	[23]
Agriculture		biodegradability, non-toxicity, antibacterial, cells activator, disease and insect resistant ability	[24]
textile industry		microorganism resistance, absorption of anionic dyes	[25]
paper finishing		high density of positive charge, non-toxicity, biodegradability, biocompatibility, antimicrobial and antifungal	[26]
solid-state batteries		ionic conductivity	[27]
chromatographic separations		the presence of free -NH2, primary -OH, secondary -OH	[28]
chitosan gel for LED and NLO applications		dye containing chitosan gels	[29]

2.2. Nanobiocomposites with Chitosan Matrix

Chitosan biopolymers have a great potential in biomedical applications due to their biocompatibility and biodegradability properties. However, the low physical properties of chitosan are most important challenge that has limited their applications. The development of high-performance chitosan biopolymers has received incorporating fillers that display a significant mechanical reinforcement [30].

Polymer nanocomposites are polymers that have been reinforced by nanosized particles with high surface area to volume ratio including nanoparticles, nanoplatelet, nanofibers, and carbon nanotubes. Nowadays, carbon nanotubes are considered to be highly potential fillers due to improving the materials properties of biopolymers [31]. Following these reports, researchers assessed the effect of CNTs fillers in chitosan matrix. Results of these research studies showed appropriate properties of CNTs/chitosan nanobiocomposites with high potential of biomedical science.

3. CARBON NANOTUBES

The carbon nanotube, which is a tubular of Buckminster fullerene, was first discovered by Iijima in 1991 [32]. These are straight segments of tube with arrangements of carbon hexagonal units [33-34]. Scientists have greatly attended to CNTs during recent years due to the existence of superior electrical, mechanical and thermal properties [35]. Carbon nanotubes are classified as single-walled carbon nanotubes (SWNTs) formed by a single graphene sheet, and multi-walled carbon nanotubes (MWNTs) formed by several graphene sheets that have been wrapped around the tube core [36]. The typical range of diameters of carbon nanotubes are a few nanometers (~0.8-2 nm at SWNTs [37-38] and ~10-400 nm at MWNTs [39]) , and their lengths are up to several micrometers [40].There are three significant methods for synthesizing CNTs including arc-discharge [41], laser ablation [42], and chemical vapor deposition (CVD) [43]. The production of CNTs also can be realized by other synthesis techniques such as the substrate [44] the sol-gel [45], and gas phase metal catalyst [46].

The C$-$C covalent bonding between the carbon atoms are similar to graphite sheets formed by sp^2 hybridization. As the result of this structure, CNTs exhibit a high specific surface area (about 10^3) [47] and thus a high tensile strength (more than 200 GPa) and elastic modulus (typically 1-5 TPa)

[48]. Carbon nanotubes have also very high thermal and electrical conductivity. However, these properties are different in a variety of employed synthesis methods, defects, chirality, the degree of graphitization, and diameter [49]. For instance, the CNT can be metallic or semiconducting, depending on the chirality [50].

Preparation of CNTs solution is impossible due to their poor solubility. Also, a strong van der waals interaction of CNTs between several nanotubes leads to aggregation into bundle and ropes [51]. Therefore, the various chemical and physical modification strategies will be necessary for improving their chemical affinity [52]. There are two approaches to the surface modification of CNTs including the covalent (grafting) and non-covalent bonding (wrapping) of polymer molecule onto the surface of CNTs [53]. In addition, the reported cytotoxic effects of CNTs *in vitro* may be mitigated by chemical surface modification [54]. On the other hand, studies show that the end-caps on nanotubes are more reactive than sidewalls. Hence, adsorption of polymers onto surface of CNTs can be utilized together with functionalization of defects and associated carbons [55].

The chemical modification of CNTs by covalent bonding is one of the important methods for improving their surface characteristics. Because of the extended π-network of the sp^2-hybridized nanotubes, CNTs have a tendency for covalent attachment, which introduces the sp^3-hibrydized C atoms [56]. These functional groups can be attached to termini of tubes by surface-bound carboxylic acids (grafting to) or direct sidewall modifications of CNTs that are based on the "in situ polymerization processing" (grafting from) [57]. Chemical functionalization of CNTs creates various activated groups (such as carboxyl [58], amine [59], fluorine [60], etc.) onto the CNTs surface by covalent bonds. However, there are two disadvantages to these methods. Firstly, the CNT structure may be decomposed due to functionalization reaction [61] and long ultrasonication process [62]. The disruption of π electron system is reduced as a result of these damages, leading to reduction of electrical and mechanical properties of CNTs. Secondly, the acidic and oxidation treatments that are often used for the functionalization of CNTs are environmentally unfriendly [63]. Thus, non-covalent functionalization of CNTs is greatly attended because of preserving their intrinsic properties while improving solubility and processability. In this method, non-covalent interaction between the π electrons of sp^2 hybridized structure at sidewalls of CNTs and other π electrons are formed by π-π stacking [64]. These non-covalent interactions can be raised between CNTs and amphiphilic molecules (surfactants) (Figure 3a). [65], polymers [66], and biopolymers such as DNA

[67], polysaccharides [68] etc. In the first method, surfactants including non-ionic surfactants, anionic surfactants and cationic surfactants are applied for functionalization of CNTs. The hydrophobic parts of surfactants are adsorbed onto the nanotubes surface, and hydrophilic parts interact with water [69]. Polymers and biopolymers can functionalize CNTs by using two methods including endohedral (Figure 3b) and wrapping (Figure 3c). Endohedral method is a strategy for the functionalization of CNTs. In this method, nanoparticles such as proteins and DNA are entrapped in the inner hollow cylinders of CNTs [70]. In another technique, the van der waals interactions and π-π stacking between CNTs and polymer lead to the wrapping of polymer around the CNTs [71]. Various polymers and biopolymers such as polyaniline [72], DNA [73], and chitosan [74] interact physically through wrapping of nanotube surface and π-π stacking by solubilized polymeric chain. However, Jian et al. (2002) created a technique for the non-covalent functionalization of SWNTs most similar to π-π stacking by PPE without polymer wrapping [75].

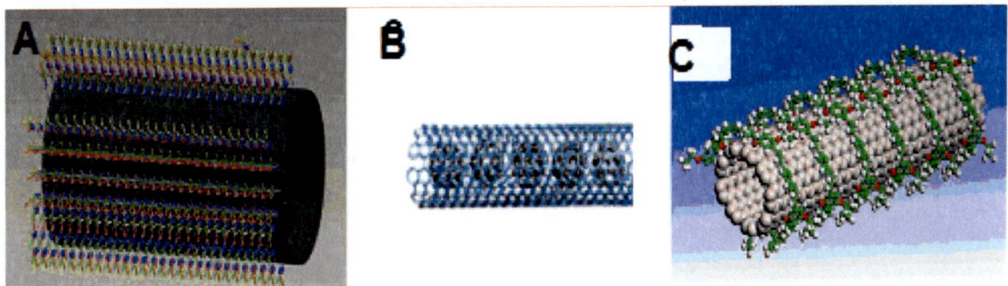

Figure 3. non-covalent functionalization of CNTs by (a) surfactants, (b) wrapping, (c) endohedral.

These functionalization methods can provide many applications of CNTs. In this context, one of the most important applications of CNTs is biomedical science such as biosensors [76], drug delivery [77], and tissue engineering [78].

3.1. Nanotube Composites

According to low physical properties of biopolymers, researchers would use some filler for the reinforcement of their electrical, mechanical, and thermal properties. Following discovery of CNTs, they have made many efforts to apply CNTs as filler in other polymers for improving properties of matrix polymer same to bulk materials [79]. The first time, Ajayan, in 1994, applied CNTs as filler in epoxy resin by the alignment method [80]. Later, many studies have focused on CNTs as excellent substitute for conventional nanofillers in the nanocomposites. Recently, many polymers and biopolymers have been reinforced by CNTs. As mentioned earlier, these nanocomposites have remarkable characteristics compared with bulk materials due to their unique properties [81].

There are several parameters affecting the mechanical properties of composites including proper dispersion, large aspect ratio of filler, interfacial stress transfer, well alignment of reinforcement, and solvent choice [82].

Uniformly and stability of nanotube dispersion in polymer matrixes are most important parameters for performance of composite. Good dispersion leads to efficient load transfer concentration centers in composites and uniform stress distribution [83]. Pemg-Cheng Ma et al. reviewed dispersion and functionalization techniques of carbon nanotubes for polymer-based nanocomposites and their effects on the properties of CNT/polymer nanocomposites. They demonstrated that the control of these two factors lead to uniform dispersion. Overall, the result showed that the proper dispersion enhanced a variety of mechanical properties of nanocomposites [71].

Fiber aspect ratio, defined as "the ratio of average fiber length to fiber diameter." This parameter is one of the main effective parameters on the longitudinal modulus [84]. Carbon nanotubes generally have high aspect ratio but their ultimate performance in a polymer composite is different. The high aspect ratio of dispersed CNTs could lead to a significant load transfer [85]. However, aggregation of the nanotubes could lead to decrease of effective aspect ratio of the CNTs. Hence, properties of nanotube composites are lower enhanced than predictions. This is one of the processing challenges and poor CNTs dispersion [86].

The interfacial stress transfer has been performed by employing external stresses to the composites. The assessments showed that fillers take a significantly larger share of the load due to CNTs-polymer matrix interaction. Also, the literature on mechanical properties of polymer nanotube composites **represented enhancement of Young's modulus** due to adding CNTs [87]. Wagner et al. investigated the effect of stress-induced fragmentation of multi-walled carbon nanotubes in a polymer matrix. The results showed that polymer deformation generates tensile stress and then transmits to CNTs [88].

The alignment CNT/polymer matrix in composite homogeneously is another effective parameter in properties of carbon nanotube composites. Quin Wang et al. [89], for instance, assessed the effects of CNT alignment on electrical conductivity and mechanical properties of SWNT/epoxy nanocomposites. The electrical conductivity, **Young's modulus** and tensile strength of the SWNT/ epoxy composite rise with increasing SWNT alignment due to increase of interface bonding of CNTs in the polymer matrix.

Umar Khan et al. in 2007, examined the effect of solvent choice on the mechanical properties of CNTs–polymer composites. They were fabricated double-walled nanotubes and polyvinyl alcohol composites into the different solvents including water, DMSO and NMP. This work shows that solvent choice can have a dramatic effect on the mechanical properties of CNTs-polymer composites [90]. Also, critical CNTs concentration has defined as optimum improvement of mechanical properties of nanotube composites where a fine network of filler formed [91].

There are other effective parameters in mechanical properties of nanotube composite such as size, crystallinity, crystalline orientation, purity, entanglement, and straightness. Generally, the ideal CNT properties depend on matrix and application [92].

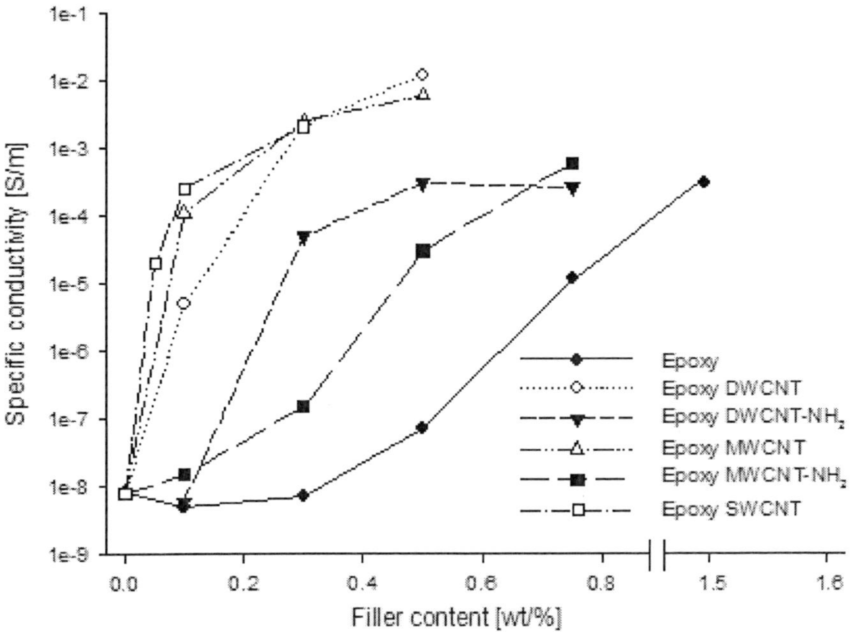

Figure 4. Electrical conductivity of the nanocomposites as function of filler content in weight percent [97].

The various functional groups on CNTs surface enable coupling with polymer matrix. A strong interface between coupled CNT/polymer creates efficient stress transfer. As a previous point, stress transfer is a critical parameter for control of mechanical properties of composite. However, covalent treatment of CNT reduces electrical [93] and thermal [94-95] properties of CNTs. These reductions affect final properties of nanotubes.

Matrix polymer can wrap around CNT surface by non-covalent functionalization. This process causes improvement in composite properties through various specific interactions. These interactions can improve properties of nanotube composites [96]. In this context, Gojny et al. [97] evaluated electrical and thermal conductivity in CNTs/epoxy composites. Figure 4 and 5 Show, respectively, electrical and thermal conductivity in various filler content including carbon black (CB), double-walled carbon nanotube (DWNT), and multi-functionalization. The experimental results represented that the electrical and thermal conductivity in nanocomposites improve by non-covalent functionalization of CNTs.

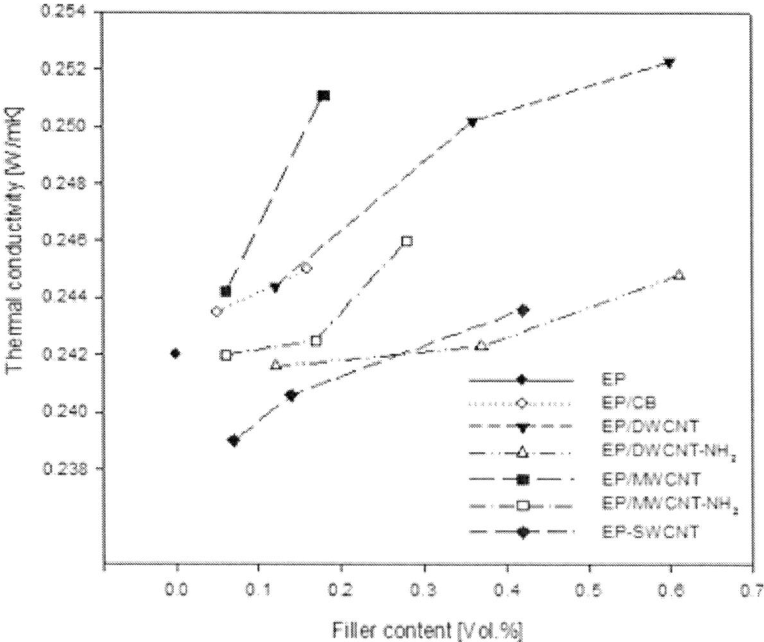

Figure 5. Thermal conductivity as function of the relative provided interfacial area per gram composite (m2/g) [97].

3.2. Mechanical and Electrical Properties of Carbon Nanotube/natural Biopolymer Composites

Table 2 represents mechanical and electrical information of CNTs/natural polymer compared with neat natural polymer. These investigations show the higher mechanical and electrical properties of CNTs/natural polymers than neat natural polymers.

Table. 2. mechanical and electrical information of neat biopolymers compared with their carbon nanotube nanocomposites

Method	Biopolymer	Mechanical					Conductivity	Ref
		Tensile modulus (Mpa)	*Tensile strength (Mpa)*	*Strain to failur (%)*	*Comparation modulus (Pa)*	*Storage modulus (Gpa)*		
Polymerized hydrogel	neat collagen				1284± 94		11.37ms± 0.16	[98]
	collagen/ CNTs				1127± 73		11.85ms± 0.67	
Solution-evaporation	neat chitosan	1.08± 0.04	37.7± 4.5				0.021 nS/cm	[99-100]
	chitosan/ CNTs	2.15± 0.09	74.3± 4.6				120 nS/cm	
Wet spinning	neat chitosan	4250						[101]
	chitosan/C NTs	10250						
Electrospinning	neat silk	140± 2.21	6.18± 0.3	5.78± 0.65			0.028 S/cm	[102]
	silk/ CNTs	4817.2 4± 69.23	44.46± 2.1	1.22± 0.14			0.144 S/cm	
Dry-jet wet spinning	neat cellulose	13100± 1100	198± 25	2.8± 0.7		5.1	negligible	[103-104]
	cellulose/ CNTs	14900± 13 00	257± 9	5.8± 1.0		7.4	3000 S/cm	
Electrospinning	neat cellulose	553± 39	21.9± 1.8	8.04± 0.27				[105]
	cellulose/ CNT	1144± 37	40.7± 2.7	10.46± 0.33				

3.3. Carbon Nanotube Composite Application

Great attention has been paid in recent years to applying nanotube composites in various fields. Wang and T.W. Yeow [106] reviewed nanotubes composites based on gas sensors. These sensors play important role for industry, environmental monitoring, biomedicine and so forth. The unique geometry, morphology, and material properties of CNTs led to applying them in gas sensors.

There are many topical studies for biological and biomedical applications of carbon nanotube composites due to its biocompatibility [107]. These components promoted biosensors [108], tissue engineering [95], and drug delivery [109] fields in biomedical technology.

On the other hand, light weight, mechanical strength, electrical conductivity, and flexibility are significant properties of carbon nanotubes for aerospace applications [110].

Kang et al. [111] presented an overview of carbon nanotube composite applications including electrochemical actuation, strain sensors, power harvesting, and bioelectronic sensors. They presented appropriate elastic and electrical properties for using nanoscale smart materials to synthesize intelligent electronic structures. In this context, Motaghitalab et al. developed polyaniline/SWNTs composite fiber [112] and showed high strength, robustness, good conductivity and pronounced electroactivity of the composite. They presented new battery materials [113] and enhancement of performance artificial muscles [114] by using these carbon nanotube composites.

Thai Ong et al. [115] addressed sustainable environment and green technologies perspective for carbon nanotube applications. These contexts include many engineering fields such as waste water treatment, air pollution monitoring, biotechnologies, renewable energy technologies, and green nanocomposites.

Sariciftici et al. [116] for the first time discovered photo-induced electron transfer from CNTs. Later, optical and photovoltaic properties of carbon nanotube composites have been studied by many groups. Results suggested the possible creation of photovoltaic devices due to hole-collecting electrode of CNTs [117].

Food packaging is another remarkable application of carbon nanotube composites. Usually, poor mechanical and barrier properties have limited applying biopolymers. Hence, appropriate filler is necessary for promotion of matrix properties. Unique properties of CNTs have been improved thermal stability, strength and modulus, and better water vapor transmission rate of applied composites in this industry [118].

4. CHITOSAN/CARBON NANOTUBE COMPOSITES

In recent decades, scientists have been interested in the creation of chitosan/CNTs composite due to providing unexampled properties of this composite. They attempted to create new properties by adding the CNTs to chitosan biopolymers. In recent years, several research articles were published in variety of applications. We summarized all of the applications of chitosan/ CNTs nanocomposites by these articles in a graph at Figure 6.

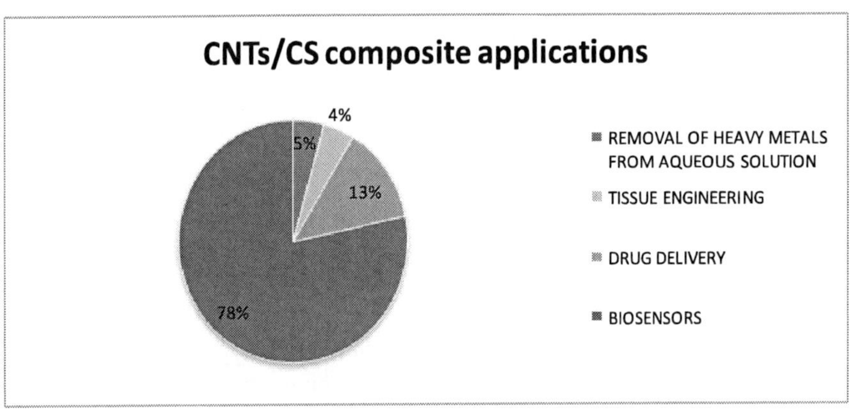

Figure 6. the graph of CNTs/CS nanocomposite application.

4.1. Chitosan/Carbon Nanotube Nanofluids

Viscosity and thermal conductivity of nanofluids containing MWNTs stabilized by chitosan were investigated by Phuoc et al. [119]. The MWNTs fluid was stabilized by chitosan solution. Studies showed that thermal conductivity enhancements obtained were significantly higher than those predicted using the Maxwell's theory. Also, they observed that dispersing chitosan into deionized water increased the viscosity of nanofluid significantly and have behaved as nan-Newtonian fluid.

4.2. Preparation Methods of CNTs/CS Nanocomposites

There are several methods for creation of nanobiocomposites. Among them, researchers have studied some of these methods for preparation the CNTs/CS nanocomposites. We represented these methods in the continuance of our review.

4.2.1. Solution-Casting-evaporation

Zhang et al. [120] assessed electrochemical sensing of carbon nanotube/chitosan system on dehydrogenase enzymes for preparing glucose biosensor first in 2004. They prepared the nanotube composite by use of solution-casting-evaporation method. In this method, the CNT/CS films were prepared by casting of CNT/CS solution on the surface of glassy carbon electrode and then drying. This CNT/CS system showed a new biocomposite platform for development of dehydrogenase-based electrochemical biosensors due to providing a signal transduction of CNT. The great results of this composite in biomedical application led to many studies in this context.

The effect of CNT/CS matrix on direct electron transfer of glucose oxidase and glucose biosensor was examined by Liu and Dong et al. [121]. They exhibited high sensitivity and better stability of CNT/CS composites compared with pure chitosan films. Furthermore, Tkac et al. [122] used the SWNT/CS films for preparation a new galactose biosensor with highly reliable detection of galactose. Tsai et al. [123] immobilized lactate dehydrogenase within MWNT/CS nanocomposite for producing lactate biosensors. This proposed biosensor provided a fast response time and high sensitivity. Also, Zhou and Chen et al. [124] showed that the immobilization of GOD molecules into chitosan-wrapped SWNT film is an efficient method for the development of a new class of very sensitive, stable, and reproducible electrochemical biosensors.

Several experiments were performed on DNA biosensor based on chitosan film doped with carbon nanotubes by Yao et al. [125]. They found that CNT/CS film can be used as a stable and sensitive platform for DNA detection. The results demonstrated improving sensor performance by adding CNT to chitosan film. Moreover, the analytical performance of glassy carbon electrodes modified with a dispersion of MWNT/CS for quantification of DNA was reported by Bollo et al. [126]. This new platform immobilized the DNA and opened the door to new strategies for development of biosensors.

In other experiments, Zeng et al. [127] reported high sensitivity of glassy carbon electrode modified by MWNT-CS for cathodic stripping voltammetric measurement of bromide (Br-).

Qian et al. [128] prepared amperometric hydrogen peroxide biosensor based on composite film of MWNT/CS. The results showed excellent electro-catalytical activity of the biosensor for H_2O_2 with good repeatability and stability.

Liu and Dong et al. [129] reported effect of CNT/CS matrix on amperometric laccase biosensor. Results showed some major advantages of this biosensor involving detecting different substrates, possessing high affinity and sensitivity, durable long-term stability, and facile preparation procedure.

Gordon Wallace et al. [130] paid particular attention to preparing of SWNT/CS film by solution-cast method and then characterizing their drug delivery properties. They found that the SWNT/CS film has enhanced slowing down release of dexamethasone.

Growth of apatite on chitosan-multi-walled carbon nanotube composite membranes at low MWNT concentrations was reported by Yang et al. [131]. Apatite was formed on the composites with low concentrations.

Immunosensors can detect various substances from bacteria to environmental pollutants. CNT/CS nanobiocomposite for immunosensor fabricated by Kaushik et al. [132]. Electron transport in this nanobiocomposite enhanced and improved the detection of ochratoxin-A, due to high electrochemical properties of SWNT. Also, CNT/CS nanocomposite used for detection of human chorionic gonadotrophin antibody was performed by Yang et al. [133] and displayed high sensitivity and good reproducibility.

4.2.1.1. Properties and Characterization

Wang et al. [134] represented that morphology and mechanical properties of chitosan has been promoted by adding CNTs. Beside, Zheng et al. [135] proved that conducting direct electron is very useful for adsorption of hemoglobin in CNT/CS composite film. These studies have demonstrated that this nanobiocomposite can used in many fields such as biosensing and biofuel cell approaches.

Tang et al. [136] evaluated water transport behavior of chitosan porous membranes containing MWNTs. They characterized two nanotube composites with low molecular weight CSP6K and high molecular weight CSP10K. Because of hollow nanochannel of MWNTs located among the pore network of chitosan membrane, the water transport results for CSP6K enhanced, when the MWNTs content is over a critical content. But, for CSP10K series membranes, the water transport rate decreased with increase of MWNTs content due to the strong compatibilizing effect of MWNTs.

CNT/CS nanocomposites were utilized by using poly(styrene sulfonic acid)-modified CNTs by Liu et al. [137] Thermal, mechanical, and electrical properties of CNT/CS composite film prepared by solution-casting have application potentials for separation membranes and sensor electrodes.

4.2.2. Crosslinking-Casting-Evaporation

In a new approach, MWNTs functionalized with –COOH groups at the end or at the sidewall defects of nanotubes by carbon nanotubes in nitric acid solvent. The functionalized carbon nanotubes immobilized into chitosan films by Emilian Ghica et al. [138]. This film applied in amperometric enzyme biosensors and resulted glucose detection and high sensitivity.

In a novel method, Kandimalla and Ju [139] cross-linked chitosan with free –CHO groups by glutaraldehyde and then MWNTs were added to the mixture. The cross-linked MWNT-CS composite immobilized acetylcholinesterase (AChE) for detecting of both acetylthiocholine and organophosphorous insecticides. On the other hand, Du et al. [140] created a new method for crosslinking CS with carboxylated CNT. This new method was performed by adding glutaraldehyde to MWNT/CS solution. They immobilized AChE on the composite for preparing an amperometric acetylthiocholine sensor. The suitable fabrication reproducibility, rapid response, high sensitivity, and stability could provide an amperometric detection of carbaryl and treazophos [141] pesticide. Results reported by Abdel Salam et al. [142] showed the removal of heavy metals including copper, zinc, cadmium, and nickel ions from aqueous solution in MWNT/CS nanocomposite film.

4.2.3. Surface Deposition Crosslinkig

Liu et al. [143] decorated carbon nanotube with chitosan by surface deposition and crosslinking process. In this new method, chitosan macromolecules as polymer cationic surfactants were adsorbed on the surface of the CNTs. In this step, CS is capable of stable dispersion of the CNT in acidic aqueous solution. The pH value of the system was increased by ammonia solution to become non-dissolvable of chitosan in aqueous media. Consequently, the soluble chitosan deposited on the surface of carbon nanotubes is similar to chitosan coating. Finally, the surface-deposited chitosan was cross-linked to the CNTs by glutaraldehyde. They found potential applications in biosensing, gene and drug delivering for this composite.

4.2.4. Electrodeposition Method

Luo and Chen et al. [144] used nanocomposite film of CNT/CS as glucose biosensor by a simple and controllable method. In this one-step electro-deposition method, a pair of gold electrodes was connected to a direct current power supply and then dipped into the CNT/CS solution. Herein, the pH near the cathode surface increased, thereby solubility of chitosan decreased. In pH of about 6.3, chitosan become insoluble and the chitosan entrapped CNT will deposited onto the cathode surface.

Yao et al. [145] also characterized electrocatalytic oxidation and sensitive electroanalysis of NADH on a novel film of CS-DA-MWNTs and improved detection sensitivity. In this new method, glutaraldehyde crosslinked CS-DA with the covalent attachment of DA molecules to CS chains formed by Schiff bases. Following, solution of MWNT dispersed in CS-DA solution dropped on an Au electrode for preparing CS-DA-MWNTs film and finally dried.

4.2.5. Covalently Grafting

Carboxylic acid (-COOH) groups were formed on the walls of CNTs by refluxing of CNTs in acidic solution. The carboxylated CNTs were added to aqueous solution of chitosan. Grafting reactions were accomplished by purging with N_2 and heated to 98 $^\circ$C of CNTs/CS solution. Shieh et al. [146] compared mechanical properties and water stability of CNTs-grafted-CS with the ungrafted CNTs. A significantly improved dispersion in chitosan matrix has resulted an important improvement storage modulus and water stability of the chitosan nanocomposites.

Wu et al. [147] created another process for make a CS-grafted MWNT composite. In this different method, after preparing oxidized MWNT (MWNT-COOH), they generated the acyl chloride functionalized MWNT (MWNT-COCl) in a solution of thionyl chloride. In the end, the MWNT-grafted-CS was synthesized by adding CS to MWNT-COCl suspension in anhydrous dimethyl formamide. The covalent modification has improved interfacial bonding and resulted high stability of CNT dispersion. Biosensors and other biological applications are evaluated as potential usage of this component. Also, Carson et al. [148] prepared a similar composite by reacting CNT-COCl and chitosan with potassium persulfate, lactic acid, and acetic acid solution at 75 $^\circ$C. They estimated that the CNT-grafted-CS composite can be used in bone tissue engineering because of the improvement of thermal properties.

4.2.5.1. Nucleophilic Substitution Reaction

Covalent modification of MWNT was accomplished with a low molecular weight chitosan (LMCS) by Ke et al. [149]. In this method, the acyl chloride functionalized grafted to LMCS in DMF/Pyridine solution. This novel derivation of MWNTs can be solved in DMF, DMAc and DMSO, but also in aqueous acetic acid solution.

4.2.6. Electrostatic Interaction

Furthermore, Baek et al. [150] synthesized CS nanoparticles-coated fMWNTs composite by electrostatic interactions between CS particles and functionalized CNT. They prepared CS nanoparticles and CS microspheres by precipitation method and crosslinking method, respectively. The electrostatic interactions between CS particles solution in distilled deionized water and the carboxylated CNTs were confirmed by changing the pH solution. Results showed same surface charges in pH 2 (both were positively charged) and pH 8 (both were negatively charged). The electrostatic interactions can be caused at pH 5.5 due to different charges between CS particles and fCNT with positive and negative surface charges, respectively. These CS particles/CNT composite materials could be utilized for potential biomedical.

Also, Zhao et al. [151] constructed SWNTs/phosphotungstic acid modified SWNTs/CS composites using phosphotungstic acid as an anchor reagent to modify SWNTs. They succeeded in using PW_{12}-modified SWNT with a negative surface charge, and on the contrary, positively charged chitosan by electrostatic interaction. These strong interfacial interactions between SWNTs and chitosan matrix presented favorable cytocompatibility for the potential use as scaffolds for bone tissue engineering.

4.2.7. Microwave Irradiation

Yu et al. [152] created a new technique for synthesis of chitosan-modified carbon nanotube by using microwave irradiation. In this technique, MWNTs solution in nitric acid were placed under microwave irradiation and dried for purification of MWNTs. A mixture of purified MWNTs and chitosan solution was reacted in the microwave oven and then centrifuged. The yield black-colored solution was adjusted at pH 8 and centrifuged for precipitation of CNT/CS composite. This facilitated technique is much more efficient than conventional methods.

4.2.8. Layer-by-layer

Wang et al. [153] characterized MWNT/CS composite rods with layer-by-layer structure were prepared via *in situ* precipitation method. Samples were prepared by coating CS solution on internal surface of a cylindrical tube and then filling with MWNT/CS solution in acetic acid. They examined morphology, mechanical, and thermal properties of this composite rod. The excellent mechanical property of these new composite rods has made potential of bone fracture internal fixation application.

4.2.8.1. Layer-by-layer Self Assembly

Xiao-bo et al. [154] produced a homogeneous multi-layer film of MWNT/CS by using layer-by-layer self assembly method. In this method, negatively charged substrates were dipped into poly (ethyleneimine) aqueous solution, MWNTs suspension, and CS solution respectively and dried at the end. In this process, both CS and PEI solution were contained NaCl for the LBL assembly. The films showed stable optical properties and were appropriate for biosensors applications.

4.2.9. Freeze-drying

Lau et al. [155] synthesized and characterized a highly conductive, porous, and biocompatible MWNT/CS biocomposite film by freeze-drying technique. This process was performed by freezing MWNT/CS dispersion into an aluminum mold and then drying. Such a composite permitted delivery of needed antibiotics with effect of increased antibiotic efficacy in a patent by Jennings et al. [156].

4.2.10. Wet-spinning

Gordon Wallace et al. [157] recently reported that chitosan is a good dispersing agent for SWNT. They also demonstrated several methods in preparing SWNT/CS macroscopic structure in the form of films, hydrogels and fibers [158]. The CNT/CS dispersion in acetic acid was spun into an ethanol:NaOH coagulation solution bath. They demonstrated increasing mechanical properties of wet spun fibers by improving dispersion [159].

4.2.11. Electrospinning

In our recent work, the chitosan(CHT)/multi-walled carbon nanotubes (MWNTs) composite nanofiber were fabricated by using electrospinning. In our experimental researches, different solvents including acetic acid 1-90%, formic acid, and TFA/DCM were tested for the electrospinning of

 A. K. Haghi

chitosan/carbon nanotube. No jet was seen upon applying the high voltage even above 25 kV by using of acetic acid 1-30% and formic acid as the solvent for chitosan/carbon nanotube. When the acetic acid 30-90% was used as the solvent, beads were deposited on the collector. Therefore, under these conditions, nanofibers were not formed.

The TFA/DCM (70:30) solvent was the only solvent that resulted in electrospinnability of chitosan/carbon nanotube. The scanning electron microscopic (Figure images showed the homogenous fibers with an average diameter of 455 nm (306-672)) were prepared with chitosan/carbon nanotube dispersion in TFA/DCM 70:30. These nanofibers have a potential for biomedical applications.

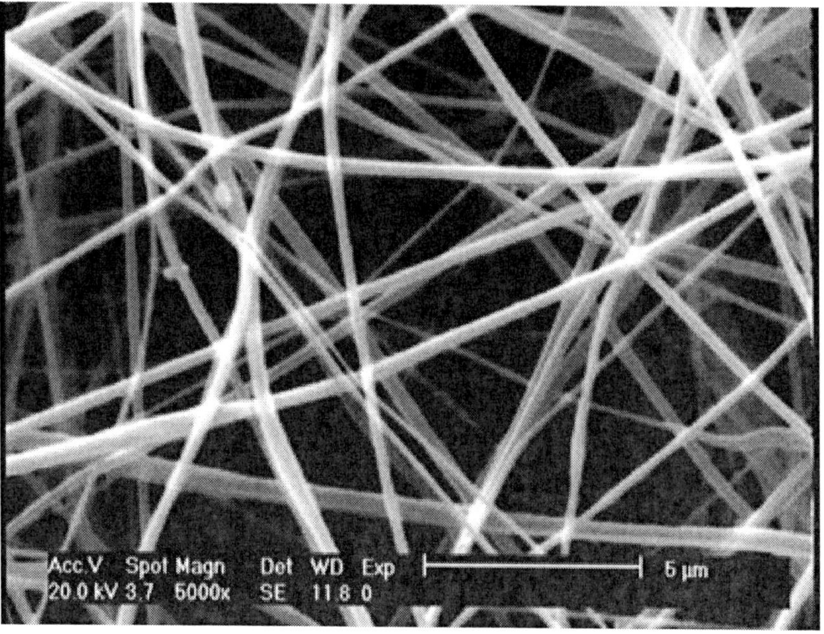

Figure 8. electron micrographs of electrospun fibers at chitosan concentration 10 wt%, 24 kV, 5 cm, TFA/DCM: 70/30.

CONCLUSION

With less than ten years history, several research studies have been created in chitosan biocomposites reinforcement using carbon nanotubes. In conclusion, much progress has been made in preparation and characterization of the CNTs/CS nanocomposites. We reported several methods for preparing these nanobiocomposites. In addition, the CNTs/CS applications have been classified including biomedicine (tissue engineering, biosensors, and drug delivery) and wastewater in this review.

Most importantly, the overriding results of electrospinning of CNTs/CS nanocomposites in our recent paper have been discussed. It is expected that there is a high potential application in tissue engineering and drug delivery by these nanobiocomposites. It is believed that with more attention to the preparation methods of CNTs/CS nanocomposites and their characterization have a promising future in biomedicine science.

REFERENCES

[1] Christian Stevens, R.V., *Renewable Bioresources: Scope and Modification for Non-food Applications.* 2004: Wiley 328.

[2] Parry, D.A.D. and E.N. Baker, Biopolymers. *Rep. Prog. Phys.,* 1984. 47: p. 1133-1232

[3] Sanjib Bhattacharyya, S.G., Hinda Dabboue, Jean-François Tranchant, Jean-Paul Salvetat*, Carbon Nanotubes as Structural Nanofibers for Hyaluronic Acid Hydrogel Scaffolds. *Biomacromolecules* 2008. 9: p. 505-509.

[4] Marino Lavorgna, F.P., Pasqualina Mangiacapra, Giovanna G. Buonocore, Study of the combined effect of both clay and glycerol plasticizer on the properties of chitosan films. *Carbohydrate Polymers,* 2010. 82(2): p. 291-298.

[5] Xiaodong Cao, Y.C., 1,2 Peter R. Chang,1 Michel A. Huneault3, Preparation and Properties of Plasticized Starch/Multi-walled Carbon Nanotubes Composites. *Journal of Applied Polymer Science,* 2007. 106: p. 1431-1437.

[6] Ramakrishna, S. et al., Biomedical applications of polymer-composite materials: a review. *Composites Science and Technology,* 2001. 61(9): p. 1189-1224.

[7] Liang, D., B.S. Hsiao, and B. Chu, Functional electrospun nanofibrous scaffolds for biomedical applications. *Advanced Drug Delivery Reviews,* 2007. 59(14): p. 1392-1412.

[8]   Zonghua Liu a, Y.J.a., Yifei Wang b, Changren Zhou a, Ziyong Zhang a,N, Polysaccharides-based nanoparticles as drug delivery systems. *Advanced Drug Delivery Reviews* 2008. 60: p. 1650–1662.

[9]   C. Agboh, O. and Y. Qin, Chitin and Chitosan Fibers. *Polymers for Advanced Technologies,* 1997. 8: p. 355–365.

[10]   Inmaculada Aranaz, M.M., Ruth Harris, Inés Paños, Beatriz Miralles, Niuris Acosta, and G.G.a.Á. Heras, Functional Characterization of Chitin and Chitosan. *Current Chemical Biology,* 2009. 3: p. 203-230.

[11]   Zhang, Y. et al., Determination of the degree of deacetylation of chitin and chitosan by X-ray powder diffraction. *Carbohydrate Research,* 2005. 340(11): p. 1914-1917.

[12]   Pamela J. VandeVord, 2 Howard W. T. Matthew,1,3 Stephen P. DeSilva,2 Lois Mayton,2 Bin Wu,2 and Paul H. Wooley1, Evaluation of the biocompatibility of a chitosan scaffold in mice. *J Biomed Mater Res,* 2002. 59(3): p. 585-590.

[13]   **Grażyna Strobin** et al., Studies on the Biodegradation of Chitosan in an Aqueous Medium. *7FIBRES & TEXTILES in Eastern Europe,* 2003. 11(3 (42)): p. 75-79.

[14]   Jayakumar, R. et al., Novel chitin and chitosan nanofibers in biomedical applications. *Biotechnology Advances,* 2010. 28(1): p. 142-150.

[15]   Bamgbose, J.T. et al., Adsorption kinetics of cadmium and lead by chitosan. *African Journal of Biotechnology,* 2010. 9(17): p. 2560-2565.

[16]   Krajewska, B., Application of chitin- and chitosan-based materials for enzyme immobilizations: a review. *Enzyme and Microbial Technology,* 2004. 35: p. 126-139.

[17]   Ueno, H., T. Mori, and T. Fujinaga, Topical formulations and wound healing applications of chitosan *Advanced Drug Delivery Reviews,* 2001. 52: p. 105-115.

[18]   Kim, I.-Y. et al., Chitosan and its derivatives for tissue engineering applications. *Biotechnology Advances,* 2008. 26: p. 1-21.

[19]   Sinha, V.R. et al., Chitosan microspheres as a potential carrier for drugs. *International Journal Of Pharmaceutics,* 2004. 274: p. 1-33.

[20]   Di Martino, A., M. Sittinger, and M.V. Risbud, Chitosan: A versatile biopolymer for orthopaedic tissue-engineering. *Biomaterials,* 2005. 26(30): p. 5983-5990.

[21]   Muzzarelli, R.A.A., Human enzymatic activities related to the therapeutic administration of chitin derivatives. *Cell. mol. life sci.,* 1997. 53: p. 131-140.

[22] Muzzarelli, R.A.A. et al., Fungistatic Activity of Modified Chitosans against Saprolegnia parasitica. *Biomacromolecules,* 2001. 2: p. 165-169.

[23] Dutta , P.K. et al., Perspectives for chitosan based antimicrobial films in food applications *Food Chemistry,* 2009. 114: p. 1173-1182.

[24] Boonlertnirun, S.B., C. Boonraung, and R. Suvanasara, Application of Chitosan in Rice Production. *Journal of Metals, Materials and Minerals* 2008. 18(2): p. 47-52.

[25] Huang , K.-S. et al., Application of low-molecular-weight chitosan in durable press finishing. *Carbohydrate Polymers,* 2008. 73: p. 254-260.

[26] Lertsutthiwong, P., S. Chandrkrachang, and W.F. Stevens, the effect of the utilization of chitosan on properties of paper. Journal of Metals, *Materials and Minerals,* 2000. 10(1): p. 43-52.

[27] Subban, R.H.Y. and A.K. Arof, Sodium iodide added chitosan electrolyte film for polymer batteries. *Physica Scripta.,* 1996. 53: p. 382-384.

[28] Ottøy, M.H. et al., Preparative and analytical size-exclusion chromatography of chitosans. *Carbohydrate Polymers,* 1996. 31(4): p. 253-261.

[29] Dutta, P.K., J. Dutta, and V.S. Tripathi, Chitin and chitosan: chemistry, properties and applications. *Journal of scientific and industrial research,* 2004. 63(20-31).

[30] Qian Li, J.Z., Lina Zhang, Structure and Properties of the Nanocomposite Films of Chitosan Reinforced with Cellulose Whiskers. *Journal of Polymer Science: Part B: Polymer Physics,* 2009? accept data. 47(11): p. 1069-1077.

[31] Thostenson, E.T., C. Li, and T.-W. Chou, Nanocomposites in context. *Composites Science and Technology,* 2005. 65(3-4): p. 491-516.

[32] Iijima, S., Helical microtubules of graphitic carbon. *Nature,* 1991. 354(6348): p. 56–58.

[33] D.S. Benthune, C.H.K., M.S. de Vries, G. Gorman, R. Savoy, J. Vazquez, R. Beyers, cobalt-catalysed growth of carbon nanotubes with single-atomic-layer-walls. *Nature,* 1993. 363: p. 605-608.

[34] Iijima, S. and T. Ichihashi, Single-shell carbon nanotubes of 1-nm diameter. *Nature,* 1993. 363: p. 603-605.

[35] Trojanowicz, M., Analytical applications of carbon nanotubes: a review. *TrAC Trends in Analytical Chemistry,* 2006. 25(5): p. 480-489.

[36] Duclaux, L., Review of the doping of carbon nanotubes (multi-walled and single-walled). *Carbon,* 2002. 40(10): p. 1751-1764.

[37] Wang, Y.Y. et al., Growth and field emission properties of small diameter carbon nanotube films. *Diamond and Related Materials.* 14(3-7): p. 714-718.

[38] Guo, J., S. Datta, and M. Lundstrom, A Numerical Study of Scaling Issues for Schottky-Barrier Carbon Nanotube Transistors. *IEEE transactions on electron devices,* 2004. 51(2): p. 172-177.

[39] Kuo, C.-S. et al., Diameter control of multi-walled carbon nanotubes using experimental strategies. *Carbon,* 2005. 43(13): p. 2760-2768.

[40] R.L. Jacobsen, T.M.T., J. R. Guth, A.C. Ehrlich, D.J. Gillespie, Mechanical properties of vapor-grown carbon fiber. *Carbon,* 1995. 33(9): p. 1217-1221.

[41] Journet C, M.W., Bernier P, Loiseau A, delaChapelle ML, Lefrant S, P.Deniard, R.Lee, J.E.Fischer, Large-scale production of single-walled carbon nanotubes by the electric-arc technique. *Nature,* 1997. 388(6644): p. 756-8.

[42] Thess A, L.R., Nikolaev P, Dai HJ, Petit P, Robert J, Chunhui Xu, Young Hee Lee, Seong Gon Kim, Andrew G. Rinzler, Daniel T, Colbert, Gustavo E, Scuseria, David Tomanek, John E. Fischer, Richard E. Smalley*, Crystalline ropes of metallic carbon nanotubes. *Science,* 1996. 273(5274): p. 483-7.

[43] Cassell AM, R.J., Kong J, Dai HJ, Large Scale CVD Synthesis of Single-Walled Carbon Nanotubes. *J Phys Chem B* 1999. 103(31): p. 6482-92.

[44] Fan S., et al., Carbon nanotube arrays on silicon substrates and their possible application. *Physica E: Low-dimensional Systems and Nanostructures,* 2000. 8(2): p. 179-183.

[45] Xie, S. et al., Carbon nanotube arrays. *Materials Science and Engineering A,* 2000. 286(1): p. 11-15.

[46] Z. K. Tang, L.Z., N. Wang, X. X. Zhang, G. H. Wen, G. D. Li, J. N. Wang, C. T. Chan, Ping Sheng*, Superconductivity in 4 Angstrom Single-Walled Carbon Nanotubes. *Science,* 2001. 292(2462-65).

[47] A. Peigney, C.L., E. Flahaut, R.R. Bacsa, A. Rousset, Specific surface area of carbon nanotubes and bundles of carbon nanotubes. *Carbon,* 2001. 39: p. 507-514.

[48] Z. W. Pan, S.S.X., a) L. Lu, B. H. Chang, L. F. Sun, W. Y. Zhou, G. Wang, and D. L. Zhang, Tensile tests of ropes of very long aligned multi-wall carbon nanotubes. *Applied Physics Letters,* 1999. 74(21): p. 3152-54.

[49] Forro, L. et al., Electronic and mechanical properties of carbon nanotubes. *Science and Application of Nanotubes,* 2000: p. 297-320.

[50] Stefan Frank, P.P., Z. L. Wang, Walt A. de Heer*, Carbon Nanotube Quantum Resistors. *Science,* 1998. 280: p. 1744-46.

[51] Khlobystov*c, D.A.B.a.A.N., Noncovalent interactions of molecules with single walled carbon nanotubes. *Chem. Soc. Rev.,* 2006. 35: p. 637-659.

[52] Andrews R, W.M., Carbon nanotube polymer composites. *Curr Opin Solid State Mater Sci* 2004. 8: p. 31-7.

[53] Hirsch*, A., Functionalization of Single-Walled Carbon Nanotubes. *Angew. Chem. Int. Ed.,* 2002. 41(11): p. 1853-59.

[54] Firme Iii, C.P. and P.R. Bandaru, Toxicity issues in the application of carbon nanotubes to biological systems. Nanomedicine: Nanotechnology, *Biology and Medicine,* 2010. 6(2): p. 245-256.

[55] S. Niyogi, M.A.H., H. Hu, B. Zhao, P. Bhowmik, R. Sen, M. E. Itkis, and R. C. Haddon*, Chemistry of Single-Walled Carbon Nanotubes. *Acc. Chem. Res.,* 2002. 35: p. 1105-1113.

[56] H. Kuzmany a, A. Kukovecz b, F. Simona, M. Holzweber a, Ch. Kramberger a, T. Pichler c, Functionalization of carbon nanotubes. *Synthetic Metals,* 2004. 141: p. 113-122.

[57] Spitalsky, Z. et al., Carbon nanotube-polymer composites: Chemistry, processing, mechanical and electrical properties. *Progress in Polymer Science,* 2010. 35(3): p. 357-401.

[58] Ravin Narain, A.H., Lynsay Lane, Modification of Carboxyl-Functionalized Single-Walled Carbon Nanotubes with Biocompatible, Water-Soluble Phosphorylcholine and Sugar-Based Polymers: Bioinspired Nanorods. *J Polym Sci A Polym Chem,* 2006. 44: p. 6558–68.

[59] Wang, M., K.P. Pramoda, and S.H. Goh, Enhancement of interfacial adhesion and dynamic mechanical properties of poly(methyl methacrylate)/multi-walled carbon nanotube composites with amine-terminated poly(ethylene oxide). *Carbon,* 2006. 44(4): p. 613-617.

[60] Touhara H, I.J., Mizuno T. et al., Fluorination of cup-stacked carbon nanotubes, structure and properties. *Fluorine Chem* 2002. 114: p. 181-8.

[61] Wei Zhang, J.K.S., Minglin Ma, Emily Y. Tsui, Stefanie A. Sydlik, Gregory C. Rutledge, Timothy M. Swager, Modular Functionalization of Carbon Nanotubes and Fullerenes. *J. Am. Chem. Soc.,* 2009. 131: p. 8446–8454.

[62] Peng He, Y.G., Jie Lian, Lumin Wang, Dong Qian, Jian Zhao, Wei Wang, Mark J. Schulz, Xing Ping Zhou, Donglu Shi, Surface modification and ultrasonication effect on the mechanical properties of carbon nanofiber/polycarbonate composites. *Composites Part A: Applied Science and Manufacturing*, 2006. 37(9): p. 1270-1275.

[63] Abu Bakar Sulong, C.H.A., Rozli Zulkifli, Mohd Roslee Othman, Joohyuk Park, A Comparison of Defects Produced on Oxidation of Carbon Nanotubes by Acid and UV Ozone Treatment. *European Journal of Scientific Research*, 2009. 33(2): p. 295-304.

[64] Wang, C. et al., Polymers containing fullerene or carbon nanotube structures. *Progress in Polymer Science*, 2004. 29(11): p. 1079-1141.

[65] Rausch, J., R.-C. Zhuang, and E. Mäder, Surfactant assisted dispersion of functionalized multi-walled carbon nanotubes in aqueous media. *Composites Part A: Applied Science and Manufacturing*, 2010. 41(9): p. 1038-1046.

[66] Sahoo, N.G. et al., Polymer nanocomposites based on functionalized carbon nanotubes. *Progress in Polymer Science*, 2010. 35(7): p. 837-867.

[67] Zheng, D., X. Li, and J. Ye, Adsorption and release behavior of bare and DNA-wrapped-carbon nanotubes on self-assembled monolayer surface. *Bioelectrochemistry*, 2009. 74(2): p. 240-245.

[68] Xiaoke Zhang, L.M., Qinghua Lu, Cell Behaviors on Polysaccharide-Wrapped Single-Wall Carbon Nanotubes: A Quantitative Study of the Surface Properties of Biomimetic Nanofibrous Scaffolds. *ACS Nano*, 2009. 10: p. 3200-6

[69] Wang, H., Dispersing carbon nanotubes using surfactants. *Current Opinion in Colloid & Interface Science*, 2009. 14(5): p. 364-371.

[70] Schlüter, A.D., A. Hirsch, and O. Vostrowsky, Functionalization of Carbon Nanotubes, in *Functional Molecular Nanostructures*. 2005, Springer Berlin / Heidelberg. p. 193-237.

[71] Peng-Cheng Ma, N.A.S., Gad Marom, Jang-Kyo Kim, Dispersion and functionalization of carbon nanotubes for polymer-based nanocomposites: A review. *Composites Part A: Applied Science and Manufacturing*, 2010. 41(10): p. 1345-1367.

[72] Mottaghitalab, V., G.M. Spinks, and G.G. Wallace, The influence of carbon nanotubes on mechanical and electrical properties of polyaniline fibers. *Synthetic Metals*, 2005. 152(1-3): p. 77-80.

[73] Cheung, W. et al., DNA and carbon nanotubes as medicine. *Advanced Drug Delivery Reviews* 2010. 62: p. 633–649.

[74] Sara Piovesan, P.A.C., James R. Smith, Dimitrios G. Fatouros and Marta Roldo, Novel biocompatible chitosan decorated single-walled carbon nanotubes (SWNTs) for biomedical applications: theoretical and experimental investigations. *Physical Chemistry Chemical Physics,* 2010. 12(48): p. 15636-15643.

[75] Jian Chen, † Haiying Liu,‡ Wayne A. Weimer,† Mathew D. Halls,† David H. Waldeck,‡ and and G.C. Walker‡, Noncovalent Engineering of Carbon Nanotube Surfaces by Rigid, Functional Conjugated Polymers. *J. AM. CHEM. SOC.,* 2002. 124: p. 9034-35.

[76] Vamvakaki, V., M. Fouskaki, and N. Chaniotakis, Electrochemical Biosensing Systems Based on Carbon Nanotubes and Carbon Nanofibers. *Analytical Letters,* 2007. 40: p. 2271-87.

[77] Kang, Y. et al., On the spontaneous encapsulation of proteins in carbon nanotubes. *Biomaterials,* 2009. 30(14): p. 2807-2815.

[78] Harrison, B.S. and A. Atala, Carbon nanotube applications for tissue engineering. *Biomaterials,* 2007. 28(2): p. 344-353.

[79] Winey*, M.M.a.K.I., Polymer Nanocomposites Containing Carbon Nanotubes. *Macromolecules,* 2006. 39: p. 5194-5205.

[80] P. M. Ajayan, O.S., C. Colliex, D. Trauth, Aligned carbon nanotube arrays formed by cutting a polymer resin-nanotube composite. *Science,* 1994. 265: p. 1212-1214.

[81] Liu, P., Modifications of carbon nanotubes with polymers. *European Polymer Journal,* 2005. 41(11): p. 2693-2703.

[82] Manchado, M.A.L. et al., Thermal and mechanical properties of single-walled carbon nanotubes-polypropylene composites prepared by melt processing. *Carbon,* 2005. 43(7): p. 1499-1505.

[83] Varadan2, Q.a.V.K., Stability analysis of carbon nanotubes via continuum models. *Smart Mater. Struct.* , 2005. 14: p. 281-286.

[84] Materials Information, E., *Fiber Reinforced Composites.* 2004: CSA Journal Division.

[85] Advani, S.G., *Processing and properties of nanocomposites.* 2006: World Scientific Publishing Company

[86] C.-W. Nan, Z.S., Y. Lin, A simple model for thermal conductivity of carbon nanotube-based composites. *Chemical Physics Letters,* 2003. 375(5-6): p. 666-669.

[87] Coleman, J.N. et al., Small but strong: A review of the mechanical properties of carbon nanotube-polymer composites. *Carbon,* 2006. 44(9): p. 1624-1652.

[88] Wagner HD, L.O., Feldman Y, Tenne R., Stress-induced fragmentation of multi-wall carbon nanotubes in a polymer matrix. *Applied Physics Letters* 1998. 72(2): p. 188–90.

[89] Wang, Q. et al., The effects of CNT alignment on electrical conductivity and mechanical properties of SWNT/epoxy nanocomposites. *Composites Science and Technology*, 2008. 68(7-8): p. 1644-1648.

[90] Khan, U. et al., The effect of solvent choice on the mechanical properties of carbon nanotube-polymer composites. *Composites Science and Technology*, 2007. 67(15-16): p. 3158-3167.

[91] Allaoui, A. et al., Mechanical and electrical properties of a MWNT/epoxy composite. *Composites Science and Technology*, 2002. 62(15): p. 1993-1998.

[92] Esawi, A.M.K. and M.M. Farag, Carbon nanotube reinforced composites: Potential and current challenges. *Materials & Design*, 2007. 28(9): p. 2394-2401.

[93] K. Kamaras, M.E.I., H. Hu, B. Zhao, R. C. Haddon†, Covalent Bond Formation to a Carbon Nanotube Metal. *Science,* 2003. 301(5639): p. 1501.

[94] Shenogin, S. et al., Effect of chemical functionalization on thermal transport of carbon nanotube composites. *Applied Physics Letters*, 2004. 85(12): p. 2229-2231.

[95] Rebecca A. MacDonald, B.F.L., Gunaranjan Viswanathan, Pulickel M. Ajayan, Jan P. Stegemann, Collagen–carbon nanotube composite materials as scaffolds in tissue engineering. *Journal of Biomedical Materials Research Part A,* 2005. 74A(3): p. 489-496.

[96] Suryasarathi Bose, R.A.K., Paula Moldenaers, Assessing the strengths and weaknesses of various types of pre-treatments of carbon nanotubes on the properties of polymer/carbon nanotubes composites: *A critical review. Polymer,* 2010 51(5): p. 975-993.

[97] Gojny, F.H. et al., Evaluation and identification of electrical and thermal conduction mechanisms in carbon nanotube/epoxy composites. *Polymer,* 2006. 47(6): p. 2036-2045.

[98] Tosun, Z. and P.S. McFetridge, A composite SWNT–collagen matrix: characterization and preliminary assessment as a conductive peripheral nerve regeneration matrix. *J. Neural Eng,* 2010. 7: p. (10pp).

[99] Wang, S.-F. et al., Preparation and Mechanical Properties of Chitosan/Carbon Nanotubes Composites. *Biomacromolecules* 2005. 6: p. 3067-3072.

[100] Liu, Y.-L., W.-H. Chen, and Y.-H. Chang, Preparation and properties of chitosan/carbon nanotube nanocomposites using poly(styrene sulfonic acid)-modified CNTs. *Carbohydrate Polymers,* 2009. 76(2): p. 232-238.

[101] Spinks, G.M. et al., Mechanical properties of chitosan/CNT microfibers obtained with improved dispersion. *Sensors and Actuators B: Chemical,* 2006. 115(2): p. 678-684.

[102] Gandhi, M. et al., Post-spinning modification of electrospun nanofiber nanocomposite from Bombyx mori silk and carbon nanotubes. *Polymer,* 2009. 50: p. 1918–1924.

[103] Rahatekar, S.S. et al., Solution spinning of cellulose carbon nanotube composites using room temperature ionic liquids. *Polymer,* 2009. 50(19): p. 4577-4583.

[104] Zhang, H. et al., Regenerated-Cellulose/Multi-walled-Carbon-Nanotube Composite Fibers with Enhanced Mechanical Properties Prepared with the Ionic Liquid 1-Allyl-3-methylimidazolium Chloride. *Adv. Mater.,* 2007. 19: p. 698–704.

[105] Lu, P. and Y.-L. Hsieh, Multi-walled Carbon Nanotube (MWCNT) Reinforced Cellulose Fibers by Electrospinning. *Applied Matterials & Interfaces,* 2010. 2(8): p. 2413–2420.

[106] Wang, Y. and J.T.W. Yeow, A Review of Carbon Nanotubes-Based Gas Sensors. *Journal of Sensors* 2009. 2009: p. 1-24.

[107] Wenrong Yang1, Pall Thordarson2, J Justin Gooding3, and S.P.R.a.F. Braet1, Carbon nanotubes for biological and biomedical applications. *Nanotechnology* 2007. 18(412001): p. (12pp).

[108] Wang*, J., Carbon-Nanotube Based Electrochemical Biosensors: A Review. *Electroanalysis* 2005. 17(1): p. 7-14.

[109] Foldvari, M. and M. Bagonluri, Carbon nanotubes as functional excipients for nanomedicines: II. Drug delivery and biocompatibility issues. *Nanomedicine: Nanotechnology, Biology and Medicine,* 2008. 4(3): p. 183-200.

[110] Belluccia, S. et al., CNT composites for aerospace applications. *Journal of Experimental Nanoscience,,* 2007. 2(3): p. 193–206.

[111] Inpil Kang a, Y.Y.H.a., Jay H. Kim b, Jong Won Lee d, Ramanand Gollapudi a, et al., Introduction to carbon nanotube and nanofiber smart materials *Composites:* Part B 2006. 37: p. 382–394.

[112] Mottaghitalab, V., G.M. Spinks, and G.G. Wallace, The development and characterisation of polyaniline--single walled carbon nanotube composite fibres using 2-acrylamido-2 methyl-1-propane sulfonic acid

(AMPSA) through one step wet spinning process. *Polymer*, 2006. 47(14): p. 4996-5002.

[113] Wang, C.Y. et al., Polyaniline and polyaniline-carbon nanotube composite fibres as battery materials in ionic liquid electrolyte. *Journal of Power Sources*, 2007. 163(2): p. 1105-1109.

[114] Mottaghitalab, V. et al., Polyaniline fibres containing single walled carbon nanotubes: Enhanced performance artificial muscles. *Synthetic Metals*, 2006. 156(11-13): p. 796-803.

[115] Yit Thai Ong, A.L.A., Sharif Hussein Sharif Zein and Soon Huat Tan*, A review on carbon nanotubes in an environmental protection and green engineering perspective. *Brazilian Journal of Chemical Engineering*, 2010. 27(02): p. 227 - 242.

[116] N. S. Sariciftci, L.S., A. J. Heeger, F. Wudi, Photoinduced Electron Transfer from a Conducting Polymer to Buckminsterfullerene. *Science*, 1992. 258: p. 1474-78.

[117] Harris*, P.J.F., Carbon nanotube composites. *International Materials Reviews*, 2004. 49(1): p. 31-43.

[118] Azeredo, H.M.C.d., Nanocomposites for food packaging applications. *Food Research International*, 2009. 42(9): p. 1240-1253.

[119] Tran X. Phuoc, M.M., Ruey-Hung Chen, Viscosity and thermal conductivity of nanofluids containing multi-walled carbon nanotubes stabilized by chitosan. *International Journal of Thermal Sciences*, 2011. 50(1): p. 12-18.

[120] Zhang M, S.A., Gorski W., Carbon nanotube-chitosan system for electrochemical sensing based on dehydrogenase enzymes. *Anal Chem*, 2004. 76(17): p. 5045-50.

[121] Liu, Y. et al., The direct electron transfer of glucose oxidase and glucose biosensor based on carbon nanotubes/chitosan matrix. *Biosensors and Bioelectronics*, 2005. 21(6): p. 984-988.

[122] Tkac, J., J.W. Whittaker, and T. Ruzgas, The use of single-walled carbon nanotubes dispersed in a chitosan matrix for preparation of a galactose biosensor. *Biosensors and Bioelectronics*, 2007. 22(8): p. 1820-1824.

[123] Tsai, Y.-C., S.-Y. Chen, and H.-W. Liaw, Immobilization of lactate dehydrogenase within multi-walled carbon nanotube-chitosan nanocomposite for application to lactate biosensors. *Sensors and Actuators B: Chemical*, 2007. 125(2): p. 474-481.

[124] Zhou, Y., H. Yang, and H.-Y. Chen, Direct electrochemistry and reagentless biosensing of glucose oxidase immobilized on chitosan

wrapped single-walled carbon nanotubes. *Talanta,* 2008. 76(2): p. 419-423.

[125] Li, J. et al., DNA biosensor based on chitosan film doped with carbon nanotubes. *Analytical Biochemistry,* 2005. 346(1): p. 107-114.

[126] Soledad Bollo, a.N.F.F., b Gustavo A. Rivasb, Electrooxidation of DNA at Glassy Carbon Electrodes Modified with Multi-wall Carbon Nanotubes Dispersed in Chitosan. *Electroanalysis,* 2007. 19(7-8): p. 833 – 840.

[127] Zeng, Y. et al., Electrochemical determination of bromide at a multi-wall carbon nanotubes-chitosan modified electrode. *Electrochimica Acta,* 2005. 51(4): p. 649-654.

[128] Qian, L. and X. Yang, Composite film of carbon nanotubes and chitosan for preparation of amperometric hydrogen peroxide biosensor. *Talanta,* 2006. 68(3): p. 721-727.

[129] Liu, Y. et al., Facile preparation of amperometric laccase biosensor with multi-function based on the matrix of carbon nanotubes-chitosan composite. *Biosensors and Bioelectronics,* 2006. 21(12): p. 2195-2201.

[130] Naficy, S. et al., Modulated release of dexamethasone from chitosan-carbon nanotube films. *Sensors and Actuators A: Physical,* 2009. 155(1): p. 120-124.

[131] Yang, J. et al., Growth of apatite on chitosan-multi-wall carbon nanotube composite membranes. *Applied Surface Science,* 2009. 255(20): p. 8551-8555.

[132] Kaushik, A. et al., Carbon nanotubes -- chitosan nanobiocomposite for immunosensor. *Thin Solid Films,* 2010. 519(3): p. 1160-1166.

[133] Yang, H. et al., Electrochemically deposited nanocomposite of chitosan and carbon nanotubes for detection of human chorionic gonadotrophin. *Colloids and Surfaces B: Biointerfaces,* 2011. 82(2): p. 463-469.

[134] Shao-Feng Wang, L.S., Wei-De Zhang,Yue-Jin Tong, Preparation and Mechanical Properties of Chitosan/Carbon Nanotubes Composites. *Biomacromolecules* 2005. 6: p. 3067-3072.

[135] W. Zheng, Y.Q.C., Y.F. Zheng Adsorption and electrochemistry of hemoglobin on Chi-carbon nanotubes composite film. *Applied Surface Science,* 2008. 255(2): p. 571-573.

[136] Changyu Tang, Q.Z., KeWang, Qiang Fua, Chaoliang Zhang, Water transport behavior of chitosan porous membranes containing multi-walled carbon nanotubes (MWNTs). *Journal of Membrane Science,* 2009. 337(1-2): p. 240-247.

[137] Ying-Ling Liu, W.-H.C., Yu-Hsun Chang, Preparation and properties of chitosan/carbon nanotube nanocomposites using poly(styrene sulfonic acid)-modified CNTs. *Carbohydrate Polymers,* 2009. 76(2): p. 232-238.

[138] Mariana Emilia Ghica, R.P., Orlando Fatibello-Filho, Christopher M.A. Brett, Application of functionalised carbon nanotubes immobilised into chitosan films in amperometric enzyme biosensors. *Sensors and Actuators B: Chemical,* 2009. 142(1): p. 308-315.

[139] Ju*[a], V.B.K.a.H., Binding of Acetylcholinesterase to Multi-wall Carbon Nanotube-Cross- Linked Chitosan Composite for Flow-Injection Amperometric Detection of an Organophosphorous Insecticide. *Chem. Eur. J.* , 2006. 12: p. 1074 – 1080.

[140] Du, D. et al., An amperometric acetylthiocholine sensor based on immobilization of acetylcholinesterase on a multi-wall carbon nanotube–cross-linked chitosan composite. *Anal Bioanal Chem* 2007. 387: p. 1059–1065.

[141] Du, D. et al., Amperometric detection of triazophos pesticide using acetylcholinesterase biosensor based on multi-wall carbon nanotube-chitosan matrix. *Sensors and Actuators B: Chemical,* 2007. 127(2): p. 531-535.

[142] Mohamed Abdel Salam, M.I.M., Magdy Y.A. Abdelaal, Preparation and characterization of multi-walled carbon nanotubes/chitosan nanocomposite and its application for the removal of heavy metals from aqueous solution. *Journal of Alloys and Compounds,* 2010. 509(5): p. 2582-2587.

[143] Yuyang Liu , J.T., Xianqiong Chen, J.H. Xin Decoration of carbon nanotubes with chitosan. *Carbon,* 2005. 43(15): p. 3178-3180.

[144] Luo, X.-L. et al., Electrochemically deposited nanocomposite of chitosan and carbon nanotubes for biosensor application. *Chem. Commun.,* 2005. DOI: 10.1039/b419197h: p. 2169–2171.

[145] Bin Ge, Y.T., Qingji Xie, Ming Ma, Shouzhuo Yao, Preparation of chitosan-dopamine-multi-walled carbon nanotubes nanocomposite for electrocatalytic oxidation and sensitive electroanalysis of NADH. *Sensors and Actuators B: Chemical,* 2009. 137(2): p. 547-554.

[146] Yeong-Tarng Shieh, Y.-F.Y., Significant improvements in mechanical property and water stability of chitosan by carbon nanotubes. *European Polymer Journal,* 2006. 42(12): p. 3162-3170.

[147] Zigang Wu, W.F., Yiyu Feng, Qiang Liu, Xinhua Xu, Tohru Sekino, Akihiko Fujii, Masanori Ozaki, Preparation and characterization of

chitosan-grafted multi-walled carbon nanotubes and their electrochemical properties. *Carbon,* 2007. 45(6): p. 1212-1218.

[148] Laura Carson, C.K.-B., Melisa Stewart, Aderemi Oki, Gloria Regisford, Zhiping Luo, Vladimir I. Bakhmutov, Synthesis and characterization of chitosan-carbon nanotube composites. *Materials Letters,* 2009. 63(6-7): p. 617-620.

[149] Gang Ke, W.C.G., Chang Yu Tang, Zhen Hu, Wen Jie Guan, Dan Lin Zeng, Feng Deng, Covalent modification of multi-walled carbon nanotubes with a low molecular weight chitosan. *Chinese Chemical Letters,* 2007. 18(3): p. 361-364.

[150] Baek, S.-H., B. Kim, and K.-D. Suh, Chitosan particle/multi-wall carbon nanotube composites by electrostatic interactions. *Colloids and Surfaces A: Physicochemical and Engineering Aspects,* 2008. 316(1-3): p. 292-296.

[151] Qichao Zhao1, J.Y., Xunda Feng1, Zujin Shi3, Zigang Ge2 *, and Zhaoxia Jin1 *, A Biocompatible Chitosan Composite Containing Phosphotungstic Acid Modified Single-Walled Carbon Nanotubes. *Journal of Nanoscience and Nanotechnology,* 2010. 10: p. 1-4.

[152] Jin-Gang Yu, K.-L.H., Jin-Chun Tang, Qiaoqin Yang, Du-Shu Huang, Rapid microwave synthesis of chitosan modified carbon nanotube composites. *International Journal of Biological Macromolecules,* 2009. 44(4): p. 316-319.

[153] Zheng-ke Wanga, b., Qiao-ling Hua, b ** and Lei Caia, b, CHITOSAN AND MULTI-WALLED CARBON NANOTUBE COMPOSITE RODS*. *Chinese Journal of Polymer Science* 2010. 28(5): p. 801-806.

[154] LI Xiao-bo, J.X.-y., Electrostatic layer-by-layer assembled multi-layer films of chitosan and carbon nanotubes. *New Carbon Materials,* 2010. 25(3): p. 237-240.

[155] Cooney*, C.L.a.M.J., Conductive Macroporous Composite Chitosan-Carbon Nanotube Scaffolds. *Langmuir,* 2008. 24: p. 7004-7010.

[156] Jennings, J.A., W.O. Haggard, and J.D. Bumgardner, *Chitosan/carbon nanotube composite scaffolds for drug delivery, in patent application publication.* 2010: United States. p. 1-8.

[157] Razal, J.M., K.J. Gilmore, and G.G. Wallace, Carbon Nanotube Biofiber Formation in a Polymer-Free Coagulation Bath. *Adv. Funct. Mater.,* 2008. 18: p. 61–66.

[158] Carol Lynam, S.E.M., and Gordon G. Wallace, Carbon-nanotube biofibers. *Adv. Mater. ,* 2007. 19: p. 1244–1248.

[159] Geoffrey M. Spinks, S.R.S., Gordon G. Wallace, Philip G. Whitten, Mechanical properties of chitosan/CNT microfibers obtained with improved dispersion. *Sensors and Actuators B: Chemical*, 2006. 115(2): p. 678-684.

In: Advanced Nanotube and Nanofiber Materials   ISBN: 978-1-62081-170-2
Editors: A. K. Haghi and G. E. Zaikov   © 2012 Nova Science Publishers, Inc.

Chapter 3

THE MODERN EXPERIMENTAL AND THEORETICAL ANALYSIS METHODS OF PARTICULATE-FILLED NANOCOMPOSITES STRUCTURE

G. V. Kozlov[1], Yu. G. Yanovskii[1] and G. E. Zaikov[2]
[1]Institute of Applied Mechanics of Russian Academy of Sciences,
Moscow, Russian Federation
[2]N.M. Emanuel Institute of Biochemical Physics
of Russian Academy of Sciences,
Moscow, Russian Federation

INTRODUCTION

The modern methods of experimental and theoretical analysis of polymer materials structure and properties allow not only confirming earlier propounded hypotheses but obtaining principally new results. Let us consider some important problems of particulate-filled polymer nanocomposites, the solution of which allows advancing substantially in these materials properties understanding and prediction. Polymer nanocomposites multi-componentness (multi-phaseness) requires their structural components quantitative characteristics determination. In this aspect, interfacial regions play a particular role, since it has been shown earlier that they are the same reinforcing element in elastomeric nanocomposites as nanofiller actually [1].

Therefore the knowledge of interfacial layer dimensional characteristics is necessary for quantitative determination of one of the most important parameters of polymer composites in general – their reinforcement degree [2, 3].

The aggregation of the initial nanofiller powder particles in more or less large particles aggregates always occurs in the course of technological process of making particulate-filled polymer composites in general [4] and elastomeric nanocomposites in particular [5]. The aggregation process tells on composites (nanocomposites) macroscopic properties [2-4]. For nanocomposites nanofiller, aggregation process gains special significance, since its intensity can be the one that nanofiller particles aggregates size exceeds 100 nm – the value, which is assumed (though conditionally enough [6]) as an upper dimensional limit for nanoparticle. In other words, the aggregation process can result in the situation when primordially supposed nanocomposite ceases to be one. Therefore at present, several methods exist, which allow to suppress nanoparticles aggregation process [5, 7]. This also assumes the necessity of the nanoparticles aggregation process quantitative analysis.

It is well known [1, 2] that in particulate-filled elastomeric nanocomposites (rubbers), nanofiller particles form linear spatial structures ("chains"). At the same time in polymer composites, filled with disperse microparticles (microcomposites), particles (aggregates of particles) of filler form a fractal network, which defines polymer matrix structure (analog of fractal lattice in computer simulation) [4]. This results in different mechanisms of polymer matrix structure formation in micro- and nanocomposites. If in the first filler particles (aggregates of particles) fractal network availability results to "disturbance" of polymer matrix structure, that is expressed in the increase of its fractal dimension d_f [4], then in case of polymer nanocomposites at nanofiller contents change the value d_f is not changed and equal to matrix polymer structure fractal dimension [3]. As it has been expected, the change of the composites of the indicated classes structure formation mechanism change defines their properties, in particular, reinforcement degree [11, 12]. Therefore, nanofiller structure fractality strict proof and its dimension determination are necessary.

As it is known [13, 14], the scale effects in general are often found at different materials mechanical properties study. The dependence of failure stress on grain size for metals (Holl-Petsch formula) [15] or of effective filling degree on filler particles size in case of polymer composites [16] are examples of such effect. The strong dependence of elasticity modulus on nanofiller particles diameter is observed for particulate-filled elastomeric nanocomposites [5]. Therefore, it is necessary to elucidate the physical grounds of nano- and micromechanical behaviour scale effect for polymer nanocomposites.

At present, a disperse material wide list is known, which is able to strengthen elastomeric polymer materials [5]. These materials are very diverse on their surface chemical constitution, but particles small size is a common feature for them. On the basis of this observation, the hypothesis was offered that any solid material would strengthen the rubber at the condition, that it was in a very dispersed state and it could be dispersed in polymer matrix. Edwards [5] points out that filler particles small size is necessary and, probably, the main requirement for reinforcement effect realization in rubbers. Using modern terminology, one can say, that for rubbers reinforcement the nanofiller particles, for which their aggregation process is suppressed as far as possible, would be the most effective ones [3, 12]. Therefore, the theoretical analysis of a nanofiller particles size influence on polymer nanocomposites reinforcement is necessary.

Proceeding from the said above, the present work's purpose is the solution of the considered-above paramount problems with the help of modern experimental and theoretical techniques on the example of particulate-filled butadiene-styrene rubber.

EXPERIMENTAL

The made industrially butadiene-styrene rubber of mark SKS-30, which contains 7.0-12.3 % cis- and 71.8-72.0 % trans-bonds, with density of 920-930 kg/m^3, was used as matrix polymer. This rubber is fully amorphous one.

Fullerene-containing mineral shungite of Zazhoginsk's deposit consists of ~ 30 % globular amorphous metastable carbon and ~ 70 % high-disperse silicate particles. Besides, industrially made technical carbon of mark № 220 was used as nanofiller. The technical carbon, nano- and microshugite particles average size makes up 20, 40 and 200 nm, respectively. The indicated filler content is equal to 37 mass %. Nano- and microdimensional disperse shungite particles were prepared from industrially output material by the original technology processing. The size and polydispersity analysis of the received in milling process shungite particles was monitored with the aid of analytical disk centrifuge (CPS Instruments, Inc., USA), allowing determination with high precision size and distribution by the sizes within the range from 2 nm up to 50 mcm.

Nanostructure was studied on atomic-forced microscopes Nano-DST (Pacific Nanotechnology, USA) and Easy Scan DFM (Nanosurf, Switzerland) by semi-contact method in the force modulation regime. Atomic-force microscopy results were processed with the help of specialized software package SPIP (Scanning Probe Image Processor, Denmark). SPIP is a powerful programmes package for processing of images, obtained on SPM, AFM, STM, scanning electron microscopes, transmission electron micro-scopes, interferometers, confocal microscopes, profilometers, optical micro-scopes and so on. The given package possesses the whole functions number, which is necessary at images precise analysis, in a number of which the following ones are included:

- the possibility of three-dimensional reflecting objects obtaining distortions automatized leveling, including Z-error mistakes removal for examination of separate elements and so on;
- quantitative analysis of particles or grains, more than 40 parameters can be calculated for each found particle or pore: area, perimeter, mean diameter, the ratio of linear sizes of grain width to its height distance between grains, coordinates of grain center of mass a.a. can be presented in a diagram form or in a histogram form.

The tests on elastomeric nanocomposites nanomechanical properties were carried out by a nanointentation method [17] on apparatus Nano Test 600 (Micro Materials, Great Britain) in load wide range from 0.01 mN up to 2.0 mN. Sample indentation was conducted in ten points with interval of 30 mcm. The load was increased with constant rate up to the greatest given load reaching (for the rate 0.05 mN/s-1 mN). The indentation rate was changed in conformity with the greatest load value counting, that loading cycle should take 20 s. The unloading was conducted with the same rate as loading. In the given experiment, the "Berkovich indentor" was used with the angle at the top of 65.3° and rounding radius of 200 nm. Indentations were carried out in the checked load regime with preload of 0.001 mN.

Elasticity modulus calculation obtained in the experiment by nanoindentation course dependences of load on indentation depth (strain) in ten points for each sample at loads of 0.01, 0.02, 0.03, 0.05, 0.10, 0.50, 1.0 and 2.0 mN were processed according to Oliver-Pharr method [18].

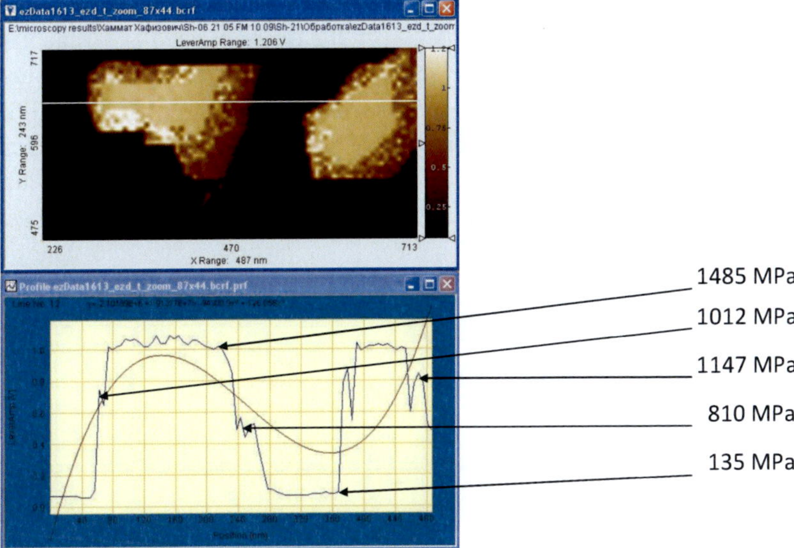

Figure 1. The processed in SPIP image of nanocomposite butadiene-styrene rubber/nanoshungite, obtained by force modulation method, and mechanical characteristics of structural components according to the data of nanoindentation (strain 150 nm).

RESULTS AND DISCUSSION

In Figure 1, presented are the obtained according to the original methodics results of elasticity moduli calculation for nanocomposite butadiene-styrene rubber/nanoshungite components (matrix, nanofiller particle and interfacial layers), received in interpolation process of nanoindentation data. The processed in SPIP polymer nanocomposite image with shungite nanoparticles allows experimental determination of interfacial layer thickness l_{if}, which is presented in Figure 1 as steps on elastomeric matrix-nanofiller boundary. The measurements of 34 such steps (interfacial layers) width on the processed in SPIP images of interfacial layer various section gave the mean experimental value l_{if}=8.7 nm. Besides, nanoindentation results (Figure 1, figures on the right) showed that interfacial layers elasticity modulus was only by 23-45% lower than nanofiller elasticity modulus, but it was higher than the corresponding parameter of polymer matrix 6.0-8.5 times. These experimental data confirm that the studied nanocomposite interfacial layer is a reinforcing element to the same extent as nanofiller actually [1, 3, 12].

Let us fulfill further the value l_{if} theoretical estimation according to the two methods and compare these results with the ones obtained experimentally. The first method simulates interfacial layer in polymer composites as a result of interaction of two fractals – polymer matrix and nanofiller surface [19, 20]. In this case, there is a sole linear scale l, which defines these fractals interpenetration distance [21]. Since nanofiller elasticity modulus is essentially higher than the corresponding parameter for rubber (in the considered case – 11 times, see Figure 1), then the indicated interaction reduces to nanofiller indentation in polymer matrix and then $l=l_{if}$. In this case, it can be written [21]:

$$l_{if} \approx a\left(\frac{R_p}{a}\right)^{2(d-d_{surf})/d} ,$$

(1)

where a is a lower linear scale of fractal behaviour, which is accepted for polymers as equal to statistical segment length l_{st} [22], R_p is a nanofiller particle (more precisely, particles aggregates) radius, which for nanoshungite is equal to ~ 84 nm [23], d is dimension of Euclidean space, in which fractal is considered (it is obvious, that in our case $d=3$), d_{surf} is fractal dimension of nanofiller particles aggregate surface.

The value l_{st} is determined as follows [24]:

$$l_{st} = l_0 C_\infty,$$

(2)

where l_0 is the main chain skeletal bond length, which is equal to 0.154 nm for both blocks of butadiene-styrene rubber [25], C_∞ is characteristic ratio, which is a polymer chain statistical flexibility indicator [26] and is determined with the help of the equation [22]:

$$T_g = 129\left(\frac{S}{C_\infty}\right)^{1/2} ,$$

(3)

where T_g is glass transition temperature, equal to 217 K for butadiene-styrene rubber [3], S is macromolecule cross-sectional area, determined for the mentioned rubber according to the additivity rule from the following considerations. As it is known [27], the macromolecule diameter quadrate values are equal: for polybutadiene $-$ 20.7 $\overset{\circ}{A}{}^2$ and for polystyrene $-$ 69.8 $\overset{\circ}{A}{}^2$. Having calculated cross-sectional area of macromolecule, simulated as a cylinder, for the indicated polymers according to the known geometrical formulas, let us obtain 16.2 and 54.8 $\overset{\circ}{A}{}^2$, respectively. Further, accepting as S the average value of the adduced above areas, let us obtain for butadiene-styrene rubber $S=35.5$ $\overset{\circ}{A}{}^2$. Then, according to the equation (3) at the indicated values T_g and S, let us obtain $C_\infty=12.5$ and according to the equation (2) $-$ $l_{st}=1.932$ nm.

The fractal dimension of nanofiller surface d_{surf} was determined with the help of the equation [3]:

$$S_u = 410 R_p^{d_{surf}-d}, \tag{4}$$

where S_u is nanoshungite particles specific surface, calculated as follows [28]:

$$S_u = \frac{3}{\rho_n R_p}, \tag{5}$$

where ρ_n is the nanofiller particles aggregate density, determined according to the formula [3]:

$$\rho_n = 0.188 \left(R_p\right)^{1/3}. \tag{6}$$

The calculation according to the equations (4)-(6) gives $d_{surf}=2.44$. Further, using the calculated by the indicated mode parameters, let us obtain from the equation (1) the theoretical value of interfacial layer thickness l_{if}^T =7.8 nm. This value is close enough to the obtained one experimentally (their discrepancy makes up $\sim$ 10 %).

The second method of value l_{if}^T estimation consists of using of the two following equations [3, 29]:

$$\varphi_{if} = \varphi_n \left(d_{surf} - 2\right) \tag{7}$$

and

$$\varphi_{if} = \varphi_n \left[\left(\frac{R_p + l_{if}^T}{R_p} \right)^3 - 1 \right],$$ (8)

where φ_{if} and φ_n are relative volume fractions of interfacial regions and nanofiller, accordingly.

The combination of the indicated equations allows receiving the following formula for l_{if}^T calculation:

$$l_{if}^T = R_p \left[\left(d_{surf} - 1 \right)^{1/3} - 1 \right].$$ (9)

The calculation according to the formula (9) gives for the considered nanocomposite $l_{if}^T = 10.8$ nm, which also corresponds well enough to the experiment (in this case discrepancy between l_{if} and l_{if}^T makes up ~ 19 %).

Let us note, in conclusion, the important experimental observation, which follows from the processed by programme SPIP results of the studied nanocomposite surface scan (Figure 1). As one can see, at one nanoshungite particle surface, from one to three (in average – two) steps can be observed, structurally identified as interfacial layers. It is significant that these steps width (or l_{if}) is approximately equal to the first (the closest to nanoparticle surface) step width. Therefore, the indicated observation supposes that in elastomeric nanocomposites at average two interfacial layers are formed: the first – at the expense of nanofiller particle surface with elastomeric matrix interaction, as a result of which molecular mobility in this layer is frozen and its state is glassy-like one; and the second – at the expense of glassy interfacial layer with elastomeric polymer matrix interaction. The most important question from the practical point of view is whether one interfacial layer or both serve as nanocomposite reinforcing element. Let us fulfill the following quantitative estimation for this question solution. The reinforcement degree (E_n/E_m) of polymer nanocomposites is given by the equation [3]:

$$\frac{E_n}{E_m} = 1 + 11 \left(\varphi_n + \varphi_{if} \right)^{1.7},$$ (10)

where E_n and E_m are elasticity moduli of nanocomposite and matrix polymer, accordingly (E_m=1.82 MPa [3]).

According to the equation (7), the sum ($\varphi_n+\varphi_{if}$) is equal to:

$$\varphi_n + \varphi_{if} = \varphi_n\left(d_{surf} - 1\right), \qquad (11)$$

if one interfacial layer (the closest to nanoshungite surface) is a reinforcing element and

$$\varphi_n + 2\varphi_{if} = \varphi_n\left(2d_{surf} - 3\right), \qquad (12)$$

if both interfacial layers are a reinforcing element.

In its turn, the value φ_n is determined according to the equation [30]:

$$\varphi_n = \frac{W_n}{\rho_n}, \qquad (13)$$

where W_n is nanofiller mass content, ρ_n is its density, determined according to the formula (6).

The calculation according to the equations (11) and (12) gave the following E_n/E_m values: 4.60 and 6.65, respectively. Since the experimental value E_n/E_m=6.10 is closer to the value calculated according to the equation (12), then this means that both interfacial layers are a reinforcing element for the studied nanocomposites. Therefore the coefficient 2 should be introduced in the equations for value l_{if} determination (for example, in the equation (1)) in case of nanocomposites with elastomeric matrix. Let us remember that the equation (1) in its initial form was obtained as a relationship with proportionality sign, i.e., without fixed proportionality coefficient [21].

Thus, the used above nanoscopic methodics allow estimating both interfacial layer structural special features in polymer nanocomposites and its sizes and properties. For the first time, it has been shown that in elastomeric particulate-filled nanocomposites, two consecutive interfacial layers are formed, which are a reinforcing element for the indicated nanocomposites. The proposed theoretical methodics of interfacial layer thickness estimation, elaborated within the frameworks of fractal analysis, give well enough correspondence to the experiment.

For theoretical treatment of nanofiller particles, aggregate growth processes and final sizes traditional irreversible aggregation models are inapplicable, since it is obvious that in nanocomposites aggregates, a large number of simultaneous growth takes place. Therefore, the model of multiple growth, offered in paper [6], was used for nanofiller aggregation description.

In Figure 2, the images of the studied nanocomposites, obtained in the force modulation regime, and corresponding to them nanoparticles aggregates fractal dimension d_f distributions are adduced. As it follows from the adduced values d_f^{ag} (d_f^{ag} =2.40-2.48), nanofiller particles aggregates in the studied nanocomposites are formed by a mechanism particle-cluster (P-Cl), i.e., they are Witten-Sander clusters [32]. The variant A was chosen, which according to mobile particles are added to the lattice, consisting of a large number of "seeds" with density of c_0 at simulation beginning [31]. Such model generates the structures, which have fractal geometry on length short scales with value $d_f \approx 2.5$ (see Figure 2) and homogeneous structure on length large scales. A relatively high particle concentration c is required in the model for uninterrupted network formation [31].

In case of "seeds" high concentration c_0 for the variant A, the following relationship was obtained [31]:

$$R_{\max}^{d_f^{ag}} = N = c / c_0, \tag{14}$$

where $R_{\max}$ is nanoparticles cluster (aggregate) greatest radius, N is nanoparticles number per one aggregate, c is nanoparticles concentration, c_0 is "seeds" number, which is equal to nanoparticles clusters (aggregates) number.

The value N can be estimated according to the following equation [8]:

$$2R_{\max} = \left(\frac{S_n N}{\pi \eta} \right)^{1/2}, \tag{15}$$

where S_n is cross-sectional area of nanoparticles, of which an aggregate consists, η is a packing coefficient, equal to 0.74 [28].

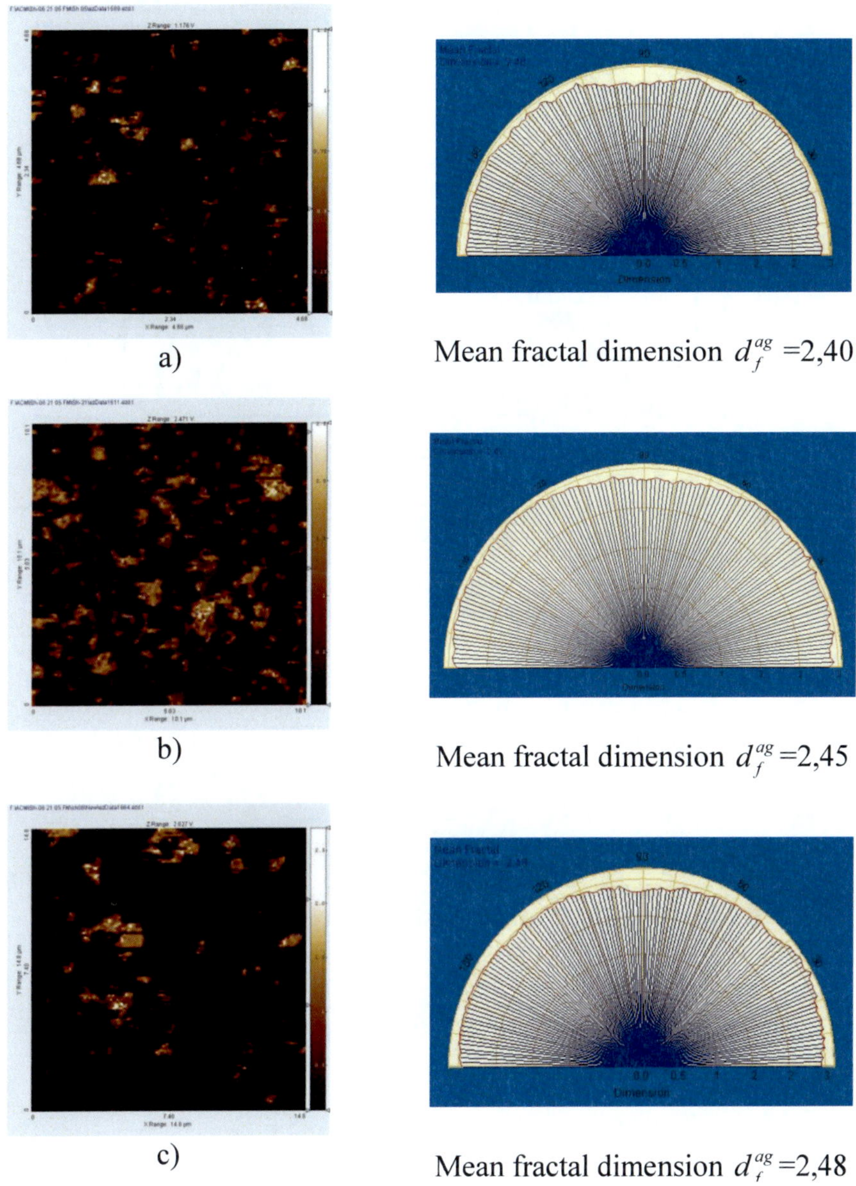

Mean fractal dimension $d_f^{ag} = 2,40$

Mean fractal dimension $d_f^{ag} = 2,45$

Mean fractal dimension $d_f^{ag} = 2,48$

Figure 2. The images, obtained in the force modulation regime, for nanocomposites, filled with technical carbon (a), nanoshungite (b), microshungite (c) and corresponding to them fractal dimensions d_f^{ag}.

G. V. Kozlov, Yu. G. Yanovskii and G. E. Zaikov

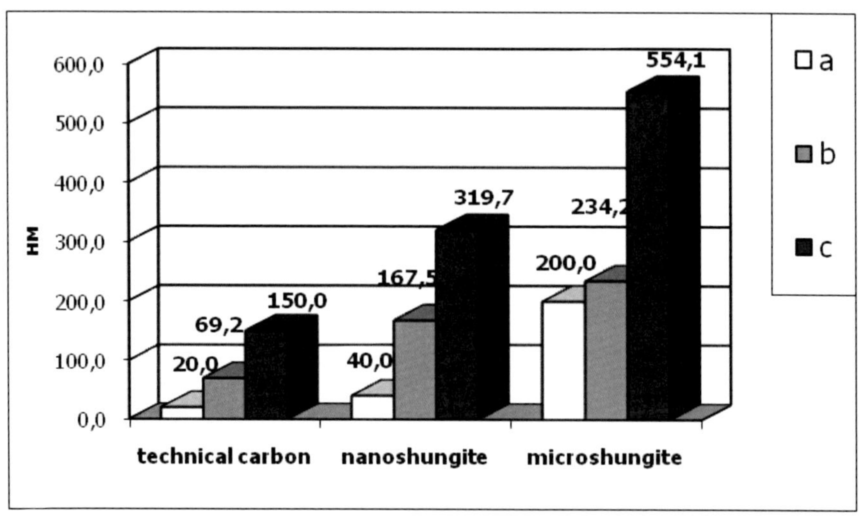

Figure 3. The initial particles diameter (a), their aggregates size in nanocomposite (b) and distance between nanoparticles aggregates (c) for nanocomposites, filled with technical carbon, nano- and microshungite.

The experimentally obtained nanoparticles aggregate diameter $2R_{ag}$ was accepted as $2R_{max}$ (Table 1), and the value S_n was also calculated according to the experimental values of nanoparticles radius r_n (Table 1). In Table 1, the values N for the studied nanofillers, obtained according to the indicated method, were adduced. It is significant that the value N is a maximum one for nanoshungite, despite larger values r_n in comparison with technical carbon.

Table 1. The parameters of irreversible aggregation model of nanofiller particles aggregates growth

Nanofiller	R_{ag}, nm	r_n, nm	N	R_{max}^{T}, nm	R_{ag}^{T}, nm	R_c, nm
Technical carbon	34.6	10	35.4	34.7	34.7	33.9
Nanoshungite	83.6	20	51.8	45.0	90.0	71.0
Microshungite	117.1	100	4.1	15.8	158.0	255.0

Further, the equation (14) allows estimating the greatest radius $R_{\max}^{T}$ of nanoparticles aggregate within the frameworks of the aggregation model [31]. These values $R_{\max}^{T}$ are adduced in Table 1, from which their reduction in a sequence of technical carbon-nanoshungite-microshungite, which fully contradicts the experimental data, i.e., to R_{ag} change (Table 1). However, we must not neglect the fact that the equation (14) was obtained within the frameworks of computer simulation, where the initial aggregating particles sizes are the same in all cases [31]. For real nanocomposites, the values r_n can be distinguished essentially (Table 1). It is expected that the higher the value R_{ag} or $R_{\max}^{T}$, the larger is the radius of nanoparticles, forming aggregate, is i.e., r_n. Then theoretical value of nanofiller particles cluster (aggregate) radius R_{ag}^{T} can be determined as follows:

$$R_{ag}^{T} = k_n r_n N^{1/d_f^{ag}},\tag{16}$$

where k_n is proportionality coefficient, in the present work accepted empirically equal to 0.9.

The comparison of experimental R_{ag} and calculated according to the equation (16) R_{ag}^{T} values of the studied nanofillers particles aggregates radius shows their good correspondence (the average discrepancy of R_{ag} and R_{ag}^{T} makes up 11.4 %). Therefore, the theoretical model [31] gives a good correspondence to the experiment only in case of consideration of aggregating particles real characteristics and, in the first place, their size.

Let us consider two more important aspects of nanofiller particles aggregation within the frameworks of the model [31]. Some features of the indicated process are defined by nanoparticles diffusion at nanocomposites processing. Specifically, length scale, connected with diffusible nanoparticle, is correlation length ξ of diffusion. By definition, the growth phenomena in sites, remote more than ξ, are statistically independent. Such definition allows connecting the value ξ with the mean distance between nanofiller particles aggregates L_n. The value ξ can be calculated according to the equation [31]:

$$\xi^2 \approx c^{-1} R_{ag}^{d_f^{ag}-d+2}, \qquad\qquad (17)$$

where c is nanoparticles concentration, which should be accepted equal to nanofiller volume contents φ_n, which is calculated according to the equations (6) and (13).

The values r_n and R_{ag} were obtained experimentally (see histogram of Figure 3). In Figure 4, the relation between L_n and ξ is adduced, which, as it is expected, proves to be linear and passing through coordinates origin. This means that the distance between nanofiller particles aggregates is limited by mean displacement of statistical walks, by which nanoparticles are simulated. The relationship between L_n and ξ can be expressed analytically as follows:

$$L_n \approx 9.6\xi ,\text{nm}. \qquad\qquad (18)$$

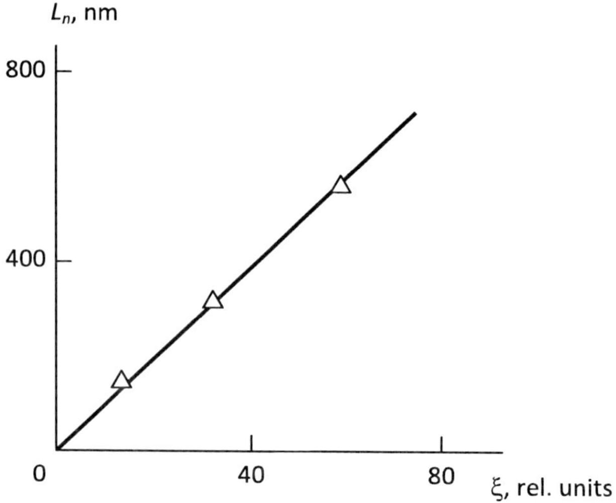

Figure 4. The relation between diffusion correlation length ξ and distance between nanoparticles aggregates L_n for considered nanocomposites.

The second important aspect of the model [31] in reference to nanofiller particles aggregation simulation is a finite nonzero initial particles concentration c or φ_n effect, which takes place in any real system. This effect is realized at the condition $\xi \approx R_{ag}$, which occurs at the critical value $R_{ag}(R_c)$, determined according to the relationship [31]:

$$c \sim R_c^{d_f^{ag}-d}. \tag{19}$$

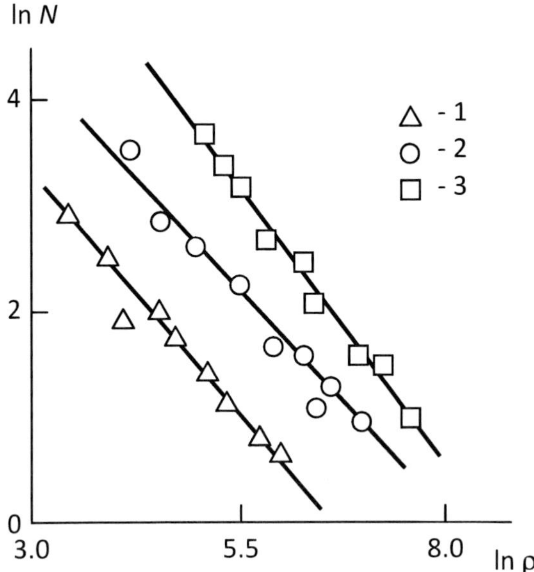

Figure 5. The dependences of nanofiller particles number N on their size ρ for nanocomposites BSR/TC (1), BSR/nanoshungite (2) and BSR/microshungite (3).

The relationship (19) right side represents cluster (particles aggregate) mean density. This equation establishes that fractal growth continues only, until cluster density reduces up to medium density, in which it grows. The calculated according to the relationship (19) values R_c for the considered nanoparticles are adduced in Table 1, from which it follows that they give reasonable correspondence with this parameter experimental values R_{ag} (the average discrepancy of R_c and R_{ag} makes up 24 %).

Since the treatment [31] was obtained within the frameworks of a more general model of diffusion-limited aggregation, then its correspondence to the experimental data indicated unequivocally that aggregation processes in these systems were controlled by diffusion. Therefore, let us consider briefly nanofiller particles diffusion. Statistical walkers diffusion constant ζ can be determined with the aid of the relationship [31]:

$$\xi \approx \left(\zeta t\right)^{1/2} \tag{20}$$

where t is walk duration.

The equation (20) supposes (at t=const) ζ increase in a number technical carbon-nanoshungite-microshungite as 196-1069-3434 relative units, i.e., diffusion intensification at diffusible particles size growth. At the same time, diffusivity D for these particles can be described by the well-known Einstein's relationship [33]:

$$D = \frac{kT}{6\pi\eta\, r_n \alpha},\tag{21}$$

where k is Boltzmann constant, T is temperature, η is medium viscosity, α is numerical coefficient, which further is accepted equal to 1.

In its turn, the value η can be estimated according to the equation [34]:

$$\frac{\eta}{\eta_0} = 1 + \frac{2.5\varphi_n}{1-\varphi_n},\tag{22}$$

where η_0 and η are initial polymer and its mixture with nanofiller viscosity, accordingly.

The calculation according to the equations (21) and (22) shows that within the indicated above nanofillers number the value D changes as 1.32-1.14-0.44 relative units, i.e., reduces in three times, which was expected. This apparent contradiction is due to the choice of the condition t=const (where t is nanocomposite production duration) in the equation (20). In real conditions, the value t is restricted by nanoparticle contact with growing aggregate and then instead of t, the value t/c_0 should be used, where c_0 is the seeds concentration, determined according to the equation (14). In this case, the value ζ for the indicated nanofillers changes as 0.288-0.118-0.086, i.e., it reduces in 3.3 times, which corresponds fully to the calculation according to Einstein's relationship (the equation (21)). This means that nanoparticles diffusion in polymer matrix obeys classical laws of Newtonian rheology [33].

Thus, the disperse nanofiller particles aggregation in elastomeric matrix can be described theoretically within the frameworks of a modified model of irreversible aggregation particle-cluster. The obligatory consideration of nano-filler initial particles size is a feature of the indicated model application to real systems description. The indicated particles diffusion in polymer matrix obeys classical laws of Newtonian liquids hydrodynamics. The offered approach allows predicting nanoparticles aggregates final parameters as a function of the initial particles size, their contents and other factors number.

At present, there are several methods of filler structure (distribution) determination in polymer matrix, both experimental [10, 35] and theoretical [4]. All the indicated methods describe this distribution by fractal dimension D_n of filler particles network. However, correct determination of any object fractal (Hausdorff) dimension includes three obligatory conditions. The first from them is the indicated above determination of fractal dimension numerical magnitude, which should not be equal to object topological dimension. As it is known [36], any real (physical) fractal possesses fractal properties within a certain scales range. Therefore, the second condition is the evidence of object self-similarity in this scales range [37]. And at last, the third condition is the correct choice of measurement scales range itself. As it has been shown in papers [38, 39], the minimum range should exceed at any rate one self-similarity iteration.

The first method of dimension D_n experimental determination uses the following fractal relationship [40, 41]:

$$D_n = \frac{\ln N}{\ln \rho},$$
(23)

where N is a number of particles with size ρ.

Particles sizes were established on the basis of atomic-power microscopy data (see Figure 2). For each, from the three studied nanocomposites, no less than 200 particles were measured, the sizes of which were united into ten groups, and mean values N and ρ were obtained. The dependences $N(\rho)$ in double logarithmic coordinates were plotted, which proved to be linear, and the values D_n were calculated according to their slope (see Figure 5). It is obvious that at such approach fractal dimension D_n is determined in two-dimensional Euclidean space, whereas real nanocomposite should be considered in three-dimensional Euclidean space. The following relationship can be used for D_n re-calculation for the case of three-dimensional space [42]:

$$D3 = \frac{d + D2 \pm \left[(d - D2)^2 - 2\right]^{1/2}}{2},$$
(24)

where $D3$ and $D2$ are corresponding fractal dimensions in three- and two-dimensional Euclidean spaces, $d=3$.

The calculated according to the indicated method dimensions D_n are adduced in Table 2. As it follows from the data of this table, the values D_n for the studied nanocomposites are varied within the range of 1.10-1.36, i.e., they characterize more or less branched linear formations ("chains") of nanofiller particles (aggregates of particles) in elastomeric nanocomposite structure. Let us remember that for particulate-filled composites polyhydroxiether/graphite, the value D_n changes within the range of ~ 2.30-2.80 [4, 10], i.e., for these materials, filler particles network is a bulk object but not a linear one [36].

Table 2. The dimensions of nanofiller particles (aggregates of particles) structure in elastomeric nanocomposites

Nanocomposite	D_n, the equation (23)	D_n, the equation (25)	d_0	d_{surf}	φ_n	D_n, the equation (29)
BSR/TC	1.19	1.17	2.86	2.64	0.48	1.11
BSR/nanoshungite	1.10	1.10	2.81	2.56	0.36	0.78
BSR/microshungite	1.36	1.39	2.41	2.39	0.32	1.47

Another method of D_n experimental determination uses the so-called "quadrates method" [43]. Its essence consists of the following. On the enlarged nanocomposite microphotograph (see Figure 2), a net of quadrates with quadrate side size α_i, changing from 4.5 up to 24 mm with constant ratio $\alpha_{i+1}/\alpha_i=1.5$, is applied and then quadrates number N_i, into which nanofiller particles hit (fully or partly), is counted up. Five arbitrary net positions concerning microphotograph were chosen for each measurement. If nanofiller particles network is a fractal, then the following relationship should be fulfilled [43]:

$$N_i \sim S_i^{-D_n/2},\qquad(25)$$

where S_i is quadrate area, which is equal to α_i^2.

In Figure 6, the dependences of N_i on S_i in double logarithmic coordinates for the three studied nanocomposites, corresponding to the relationship (25), is adduced. As one can see, these dependences are linear, which allows determining the value D_n from their slope. The determined according to the relationship (25) values D_n are also adduced in Table 2, from which a good correspondence of dimensions D_n, obtained by the two described-above methods follows (their average discrepancy makes up 2.1 % after these dimensions re-calculation for three-dimensional space according to the equation (24)).

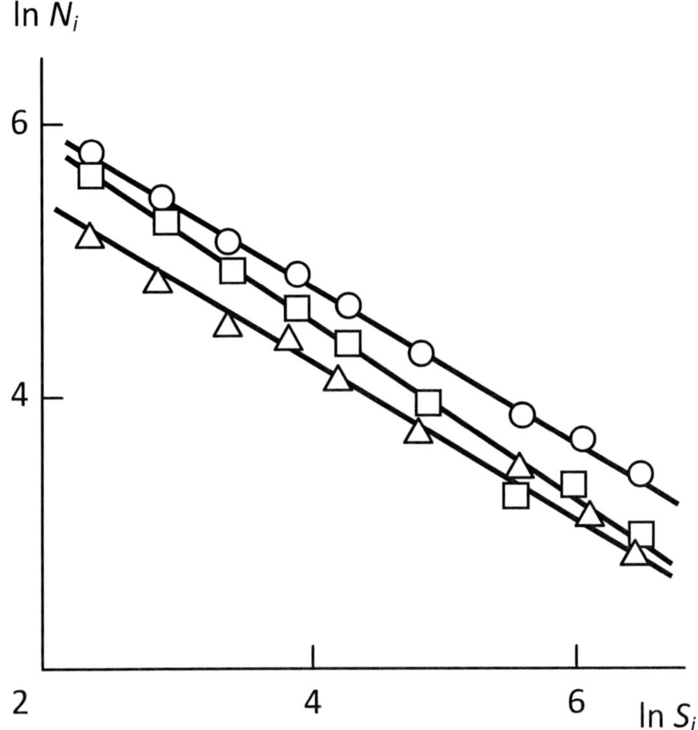

Figure 6. The dependences of covering quadrates number N_i on their area S_i, corresponding to the relationship (25), in double logarithmic coordinates for nanocomposites on the basis of BSR. The designations are the same as that in Figure 5.

As it has been shown in paper [44], the usage for self-similar fractal objects at the relationship (25) the condition should be fulfilled:

$$N_i - N_{i-1} \sim S_i^{-D_n}. \tag{26}$$

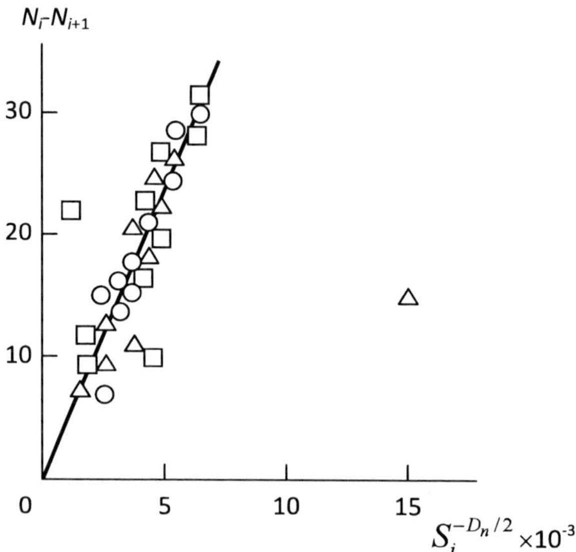

Figure 7. The dependences of (N_i-N_{i+1}) on the value $S_i^{-D_n/2}$, corresponding to the relationship (26), for nanocomposites on the basis of BSR. The designations are the same as that in Figure 5.

In Figure 7, the dependence, corresponding to the relationship (26), for the three studied elastomeric nanocomposites is adduced. As one can see, this dependence is linear, passes through coordinates origin, which according to the relationship (26) is confirmed by nanofiller particles (aggregates of particles) "chains" self-similarity within the selected α_i range. It is obvious that this self-similarity will be a statistical one [44]. Let us note that the points corresponding to α_i=16 mm for nanocomposites butadiene-styrene rubber/ technical carbon (BSR/TC) and butadiene-styrene rubber/microshungite (BSR/microshungite) do not correspond to a common straight line. Accounting for electron microphotographs of Figure 2 enlargement, this gives the self-similarity range for nanofiller "chains" of 464-1472 nm. For nanocomposite butadiene-styrene rubber/nanoshungite (BSR/nanoshungite), which has no points deviating from a straight line of Figure 7, α_i range makes up 311-1510 nm, which corresponds well enough to the indicated above self-similarity range.

In papers [38, 39], it has been shown that measurement scales S_i minimum range should contain at least one self-similarity iteration. In this case, the condition for ratio of maximum S_{max} and minimum S_{min} areas of covering quadrates should be fulfilled [39]:

$$\frac{S_{max}}{S_{min}} > 2^{2/D_n}. \tag{27}$$

Hence, accounting for the defined-above restriction, let us obtain $S_{max}/S_{min}=121/20.25=5.975$, that is larger than values $2^{2/D_n}$ for the studied nanocomposites, which are equal to 2.71-3.52. This means that measurement scales range is chosen correctly.

The self-similarity iterations number μ can be estimated from the inequality [39]:

$$\left(\frac{S_{max}}{S_{min}}\right)^{D_n/2} > 2^{\mu}. \tag{28}$$

Using the indicated above values of the included in the inequality (28) parameters, $\mu=1.42-1.75$ is obtained for the studied nanocomposites, i.e., in our experiment, conditions self-similarity iterations number is larger than unity, which again confirms correctness of the value D_n estimation [35].

And let us consider in conclusion the physical grounds of smaller values D_n for elastomeric nanocomposites in comparison with polymer microcomposites, i.e., the causes of nanofiller particles (aggregates of particles) "chains" formation in the first ones. The value D_n can be determined theoretically according to the equation [4]:

$$\varphi_{if} = \frac{D_n + 2.55d_0 - 7.10}{4.18}, \tag{29}$$

where φ_{if} is interfacial regions relative fraction, d_0 is nanofiller initial particles surface dimension.

The dimension d_0 estimation can be carried out with the help of the equation (4), and the value φ_{if} can be calculated according to the equation (7). The results of dimension D_n theoretical calculation according to the equation (29) are adduced in Table 2, from which a theory and experiment good correspondence follows. The equation (29) indicates unequivocally to the cause of a filler in nano- and microcomposites different behaviour. The high (close to 3, see Table 2) values d_0 for nanoparticles and relatively small (d_0=2.17 for graphite [4]) values d_0 for microparticles at comparable values φ_{if} is such cause for composites of the indicated classes [3, 4].

Hence, the stated-above results have shown that nanofiller particles (aggregates of particles) "chains" in elastomeric nanocomposites are physical fractal within self-similarity (and, hence, fractality [41]) range of ~ 500-1450 nm. In this range, their dimension D_n can be estimated according to the equations (23), (25) and (29). The cited examples demonstrate the necessity of the measurement scales range correct choice. As it has been noted earlier [45], the linearity of the plots, corresponding to the equations (23) and (25), and D_n nonintegral value do not guarantee object self-similarity (and, hence, fractality). The nanofiller particles (aggregates of particles) structure low dimensions are due to the initial nanofiller particles surface high fractal dimension.

In Figure 8, the histogram is adduced, which shows elasticity modulus E change, obtained in nanoindentation tests, as a function of load on indenter P or nanoindentation depth h. Since for all the three considered nanocomposites the dependences $E(P)$ or $E(h)$ are identical qualitatively, then further the dependence $E(h)$ for nanocomposite BSR/TC was chosen, which reflects the indicated scale effect quantitative aspect in the most clearest way.

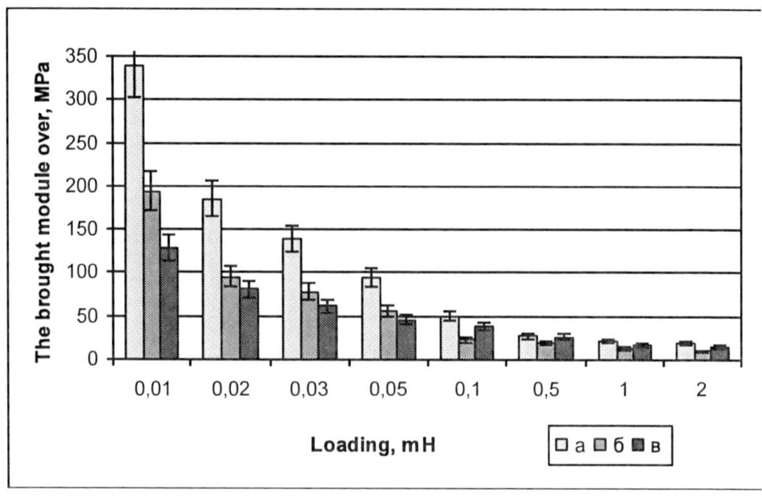

Figure 8. The dependences of reduced elasticity modulus on load on indentor for nanocomposites on the basis of butadiene-styrene rubber, filled with technical carbon (a), micro- (b) and nanoshungite (c).

In Figure 9, the dependence of E on h_{pl} (see Figure 10) is adduced, which breaks down into two linear parts. Such dependences elasticity modulus – strain are typical for polymer materials in general and are due to inter-molecular bonds anharmonicity [46]. In paper [47], it has been shown that the dependence $E(h_{pl})$ first part at $h_{pl} \leq 500$ nm is not connected with relaxation processes and has a purely elastic origin. The elasticity modulus E on this part changes in proportion to h_{pl} as:

$$E = E_0 + B_0 h_{pl},$$

(30)

where E_0 is "initial" modulus, i.e., modulus, extrapolated to $h_{pl}=0$, and the coefficient B_0 is a combination of the first and second kind elastic constants. In the considered case $B_0<0$. Further Grüneisen parameter γ_L, characterizing intermolecular bonds anharmonicity level can be determined [47]:

$$\gamma_L \approx -\frac{1}{6} - \frac{1}{2}\frac{B_0}{E_0}\frac{1}{(1-2\nu)},$$

(31)

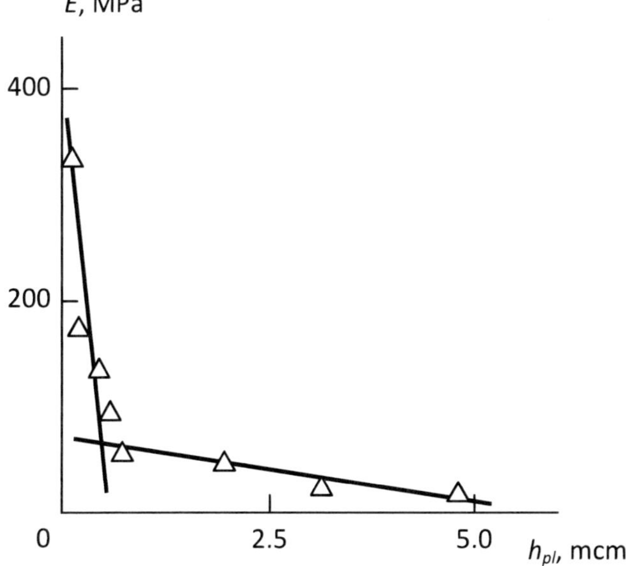

Figure 9. The dependence of reduced elasticity modulus E, obtained in nanoindentation experiment, on plastic strain h_{pl} for nanocomposites BSR/TC.

Berkovich indenter

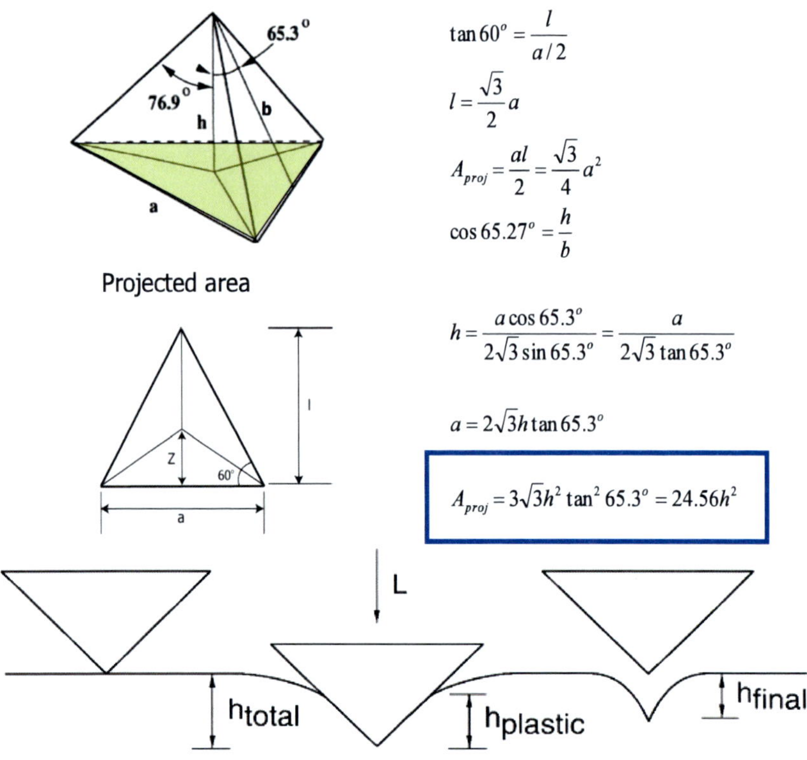

Figure 10. The schematic image of Berkovich indentor and nanoindentation process.

where ν is Poisson ratio, accepted for elastomeric materials equal to ~ 0.475 [36].

Calculation according to the equation (31) has given the following values γ_L: 13.6 for the first part and 1.50 – for the second one. Let us note the first from γ_L adduced values is typical for intermolecular bonds, whereas the second value γ_L is much closer to the corresponding value of Grüneisen parameter G for intrachain modes [46].

Poisson's ratio ν can be estimated by γ_L (or G) known values according to the formula [46]:

$$\gamma_L = 0.7\left(\frac{1+v}{1-2v}\right).$$ (32)

The estimations according to the equation (32) gave: for the dependence $E(h_{pl})$ first part $v=0.462$, for the second one - $v=0.216$. If for the first part, the value v is close to Poisson's ratio magnitude for nonfilled rubber [36], then in the second part case the additional estimation is required. As it is known [48], a polymer composites (nanocomposites) Poisson's ratio value v_n can be estimated according to the equation:

$$\frac{1}{v_n} = \frac{\varphi_n}{v_{TC}} + \frac{1-\varphi_n}{v_m},$$ (33)

where φ_n is nanofiller volume fraction, v_{TC} and v_m are nanofiller (technical carbon) and polymer matrix Poisson's ratio, respectively.

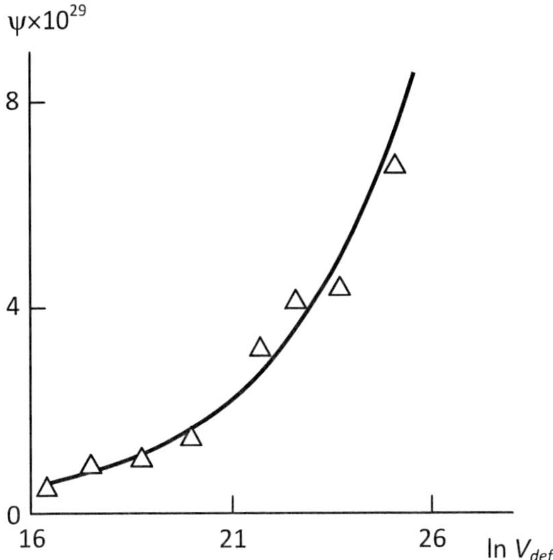

Figure 11. The dependence of density fluctuation ψ on volume of deformed in nanoindentation process material V_{def} in logarithmic coordinates for nanocomposites BSR/TC.

The value ν_m is accepted equal to 0.475 [36], and the magnitude ν_{TC} is estimated as follows [49]. As it is known [50], the nanoparticles TC aggregates fractal dimension d_f^{ag} value is equal to 2.40, and then the value ν_{TC} can be determined according to the equation [50]:

$$d_f^{ag} = (d-1)(1+\nu_{TC}).$$

(34)

According to the formula (34) ν_{TC}=0.20 and calculation ν_n according to the equation (33) gives the value 0.283, that is close enough to the value ν=0.216 according to the equation (32) estimation. The obtained by the indicated methods values ν and ν_n comparison demonstrates, that in the dependence $E(h_{pl})$ (h_{pl}<0.5 mcm) the first part in nanoindentation tests only rubber-like polymer matrix ($\nu=\nu_m\approx0.475$) is included and in this dependence the second part – the entire nanocomposite as homogeneous system [51]—$\nu=\nu_n\approx0.22$.

Let us consider further E reduction at h_{pl} growth (Figure 9) within the frameworks of density fluctuation theory, which value ψ can be estimated as follows [22]:

$$\psi = \frac{\rho_n kT}{K_T},$$

(35)

where ρ_n is nanocomposite density, k is Boltzmann constant, T is testing temperature, K_T is isothermal modulus of dilatation, connected with Young's modulus E by the relationship [46]:

$$K_T = \frac{E}{3(1-\nu)}.$$

(36)

In Figure 10, the scheme of volume of the deformed at nanoindentation material V_{def} calculation in case of Berkovich indentor using is adduced, and in Figure 11, the dependence $\psi(V_{def})$ in logarithmic coordinates was shown. As it follows from the data of this Figure, the density fluctuation growth is observed at the deformed material volume increase. The plot $\psi(\ln V_{def})$ extrapolation to ψ=0 gives $\ln V_{def}\approx13$ or $V_{def}(V_{def}^{cr})$=4.42×10^5 nm^3. Having determined the linear scale l_{cr} of transition to $\psi=0$ as $(V_{def}^{cr})^{1/3}$, let us obtain $l_{cr}=75.9$ nm, which is close to nanosystems dimensional range upper boundary (as it was noted above, conditional enough [6]), which is equal to 100 nm. Thus, the stated above results suppose that nanosystems are such systems in which density fluctuations are absent, always taking place in microsystems.

As it follows from the data of Figure 9, the transition from nano- to microsystems occurs within the range $h_{pl}=408-726$ nm. Both the indicated above values h_{pl} and the corresponding to them values $(V_{def})^{1/3}\approx814-1440$ nm can be chosen as the linear length scale l_n, corresponding to this transition. From the comparison of these values l_n with the distance between nanofiller particles aggregates L_n ($L_n=219.2-788.3$ nm for the considered nanocomposites, see Figure 3), it follows that for transition from nano- to microsystems, l_n should include at least two nanofiller particles aggregates and surrounding them layers of polymer matrix, which is the lowest linear scale of nanocomposite simulation as a homogeneous system. It is easy to see that nanocomposite structure homogeneity condition is harder than the obtained above from the criterion $\psi=0$. Let us note that such method, namely, a nanofiller particle and surrounding it polymer matrix layers separation, is widespread at a relationships derivation in microcomposite models.

It is obvious that the equation (35) is inapplicable to nanosystems, since $\psi\to0$ assumes $K_T\to\infty$, which is physically incorrect. Therefore the value E_0, obtained by the dependence $E(h_{pl})$ extrapolation (see Figure 9) to $h_{pl}=0$, should be accepted as E for nanosystems [49].

Hence, the stated above results have shown that elasticity modulus change at nanoindentation for particulate-filled elastomeric nanocomposites is due to a number of causes, which can be elucidated within the framework of anharmonicity conception and density fluctuation theory. Application of the first from the indicated conceptions assumes that in nanocomposites during nanoindentation process, local strain is realized, affecting polymer matrix only, and the transition to macrosystems means nanocomposite deformation as homogeneous system. The second from the mentioned conceptions has shown that nano- and microsystems differ by density fluctuation absence in the first and availability of ones in the second. The last circumstance assumes that for the considered nanocomposites density fluctuations take into account nanofiller and polymer matrix density difference. The transition from nano- to microsystems is realized in the case when the deformed material volume exceeds nanofiller particles aggregate and surrounding it layers of polymer matrix combined volume [49].

In work [3], the following formula was offered for elastomeric nanocomposites reinforcement degree E_n/E_m description:

$$\frac{E_n}{E_m} = 15.2\left[1 - \left(d - d_{surf}\right)^{1/t}\right],\qquad(37)$$

where t is index percolation, equal to 1.7 [28].

From the equation (37), it follows that nanofiller particles (aggregates of particles) surface dimension d_{surf} is the parameter, controlling nanocomposites reinforcement degree [53]. This postulate corresponds to the known principle about numerous division surfaces decisive role in nanomaterials as the basis of their properties change [54]. From the equations (4)-(6), it follows unequivocally that the value d_{surf} is defined by nanofiller particles (aggregates of particles) size R_p only. In its turn, from the equation (37), it follows that elastomeric nanocomposites reinforcement degree E_n/E_m is defined by the dimension d_{surf} only, or, accounting for the said above, by the size R_p only. This means that the reinforcement effect is controlled by nanofiller particles (aggregates of particles) sizes only and in virtue of this is the true nanoeffect.

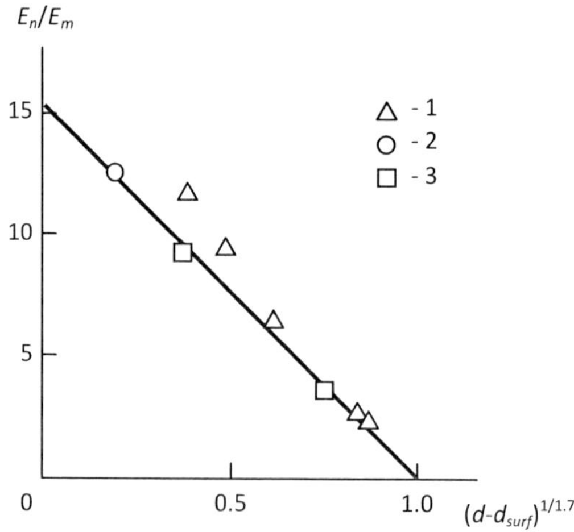

Figure 12. The dependence of reinforcement degree E_n/E_m on parameter $(d-d_{surf})^{1/1.7}$ value for nanocomposites NR/TC (1), BSR/TC (2) and BSR/shungite (3).

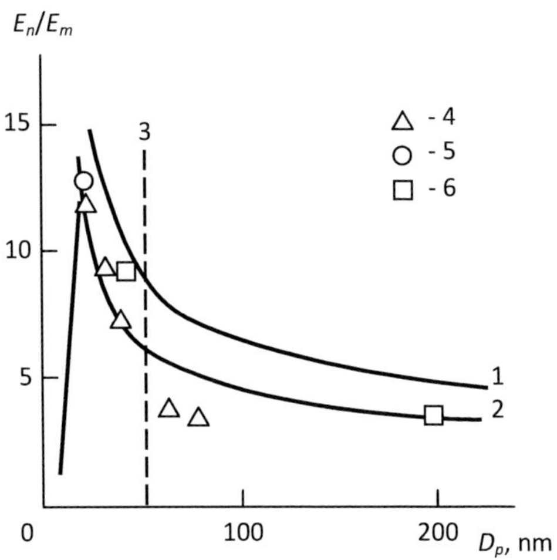

Figure 13. The theoretical dependences of reinforcement degree E_n/E_m on nanofiller particles size D_p, calculated according to the equations (4)-(6) and (37), at initial nanoparticles (1) and nanoparticles aggregates (2) size using. 3 – the boundary value D_p, corresponding to true nanocomposite. 4-6 – the experimental data for nanocomposites NR/TC (4), BSR/TC (5) and BSR/shungite (6).

In Figure 12, the dependence of E_n/E_m on $(d-d_{surf})^{1/1.7}$ is adduced, corresponding to the equation (37), for nanocomposites with different elastomeric matrices (natural and butadiene-styrene rubbers, NR and BSR, accordingly) and different nanofillers (technical carbon of different marks, nano- and microshungite). Despite the indicated distinctions in composition, all adduced data are described well by the equation (37).

In Figure 13, two theoretical dependences of E_n/E_m on nanofiller particles size (diameter D_p), calculated according to the equations (4)-(6) and (37), are adduced. However, at the curve 1 calculation, the value D_p for the initial nanofiller particles was used, and at the curve 2 calculation – nanofiller particles aggregates size D_p^{ag} (see Figure 3). As it was expected [5], the growth E_n/E_m at D_p or D_p^{ag} reduction; in addition, the calculation with D_p (nonaggregated nanofiller) using gives higher E_n/E_m values in comparison with the aggregated one (D_p^{ag} using). At $D_p \leq 50$ nm, faster growth E_n/E_m at D_p reduction is observed than at $D_p > 50$ nm, which was also expected. In Figure 13, the critical theoretical value D_p^{cr} for this transition, calculated according to the indicated above general principles [54], is pointed out by a vertical shaded line. In conformity with these principles, the nanoparticles size in nanocomposite is determined according to the condition when division surface fraction in the entire nanomaterial volume makes up about 50% and more. This fraction is estimated approximately by the ratio $3l_{if}/D_p$, where l_{if} is interfacial layer thickness. As it was noted above, the data of Figure 1 gave the average experimental value $l_{if}\approx8.7$ nm. Further, from the condition $3l_{if}/D_p\approx0.5$, let us obtain $D_p\approx52$ nm, which is shown in Figure 13 by a vertical shaded line. As it was expected, the value $D_p\approx52$ nm is a boundary one for regions of slow ($D_p>52$ nm) and fast ($D_p\leq52$ nm) E_n/E_m growth at D_p reduction. In other words, the materials with nanofiller particles size $D_p\leq52$ nm ("super-reinforcing" filler according to the terminology of paper [5]) should be considered true nanocomposites.

Let us note in conclusion that although the curves 1 and 2 of Figure 13 are similar ones, nanofiller particles aggregation, which the curve 2 accounts for, reduces essentially enough nanocomposites reinforcement degree. At the same time, the experimental data correspond exactly to the curve 2, which was to be expected in virtue of aggregation processes, which always took place in real composites [4] (nanocomposites [55]). The values d_{surf}, obtained according to the equations (4)-(6) correspond well to the determined experimentally ones. So, for nanoshungite and two marks of technical carbon, the calculation by the indicated method gives the following d_{surf} values: 2.81, 2.78 and 2.73, whereas experimental values of this parameter are equal to: 2.81, 2.77 and 2.73, i.e., practically a full correspondence of theory and experiment was obtained.

CONCLUSION

Hence, the stated-above results have shown that the elastomeric reinforcement effect is the true nanoeffect, which is defined by the initial nanofiller particles size only. The indicated particles aggregation, always taking place in real materials, changes reinforcement degree quantitatively only, namely, reduces it. This effect theoretical treatment can be received within the frameworks of fractal analysis. For the considered Nanocomposites, the nanoparticle size upper limiting value makes up ~ 52 nm.

REFERENCES

[1]   Yanovskii Yu. G., Kozlov G.V., Karnet Yu.N. *Mekhanika Kompozitsionnykh Materialov i Konstruktsii,* 2011, v. 17, № 2, p. 203-208.

[2]   Malamatov A.Kh., Kozlov G.V., Mikitaev M.A. *Reinforcement Mechanisms of Polymer Nanocomposites.* Moscow, Publishers of the D.I. Mendeleev RKhTU, 2006, 240 p.

[3]   Mikitaev A.K., Kozlov G.V., Zaikov G.E. *Polymer Nanocomposites: Variety of Structural Forms and Applications.* Moscow, Nauka, 2009, 278 p.

[4]   Kozlov G.V., Yanovskii Yu. G., Karnet Yu. N. *Structure and Properties of Particulate-Filled Polymer Composites: the Fractal Analysis.* Moscow, Al'yanstransatom, 2008, 363 p.

[5]   Edwards D.C. *J. Mater. Sci.,* 1990, v. 25, № 12, p. 4175-4185.

[6]   Buchachenko A.L. *Uspekhi Khimii,* 2003, v. 72, № 5, p. 419-437.

[7]   Kozlov G.V., Yanovskii Yu.G., Burya A.I., Aphashagova Z.Kh. *Mekhanika Kompozitsionnykh Materialov i Konstruktsii,* 2007, v. 13, № 4, p. 479-492.

[8]   Lipatov Yu. S. *The Physical Chemistry of Filled Polymers.* Moscow, Khimiya, 1977, 304 p.

[9]   Bartenev G.M., Zelenev Yu. V. *Physics and Mechanics of Polymers.* Moscow, Vysshaya Shkola, 1983, 391 p.

[10]   Kozlov G.V., Mikitaev A.K. *Mekhanika Kompozitsionnykh Materialov i Konstruktsii,* 1996, v. 2, № 3-4, p. 144-157.

[11]   Kozlov G.V., Yanovskii Yu.G., Zaikov G.E. *Structure and Properties of Particulate-Filled Polymer Composites: the Fractal Analysis.* New York, Nova Science Publishers, Inc., 2010, 282 p.

[12]   Mikitaev A.K., Kozlov G.V., Zaikov G.E. *Polymer Nanocomposites: Variety of Structural Forms and Applications.* New York, Nova Science Publishers, Inc., 2008, 319 p.

[13]   McClintok F.A., Argon A.S. *Mechanical Behavior of Materials.* Reading, Addison-Wesley Publishing Company, Inc., 1966, 440 p.

[14]   Kozlov G.V., Mikitaev A.K. *Doklady AN SSSR,* 1987, v. 294, № 5, p. 1129-1131.

[15]   Honeycombe R.W.K. *The Plastic Deformation of Metals.* Boston, Edward Arnold (Publishers), Ltd., 1968, 398 p.

[16]   Dickie R.A. In book: *Polymer Blends.* V. 1. New York, San-Francisco, London, Academic Press, 1980, p. 386-431.

[17] Kornev Yu. V., Yumashev O.B., Zhogin V.A., Karnet Yu. N., Yanovskii Yu.G. *Kautschuk i Rezina, 2008*, № 6, p. 18-23.

[18] Oliver W.C., Pharr G.M. *J. Mater. Res.,* 1992, v. 7, № 6, p. 1564-1583.

[19] Kozlov G.V., Yanovskii Yu.G., Lipatov Yu.S. *Mekhanika Kompozitsionnykh Materialov i Konstruktsii,* 2002, v. 8, № 1, p. 111-149.

[20] Kozlov G.V., Burya A.I., Lipatov Yu.S. *Mekhanika Kompozitnykh Materialov,* 2006, v. 42, № 6, p. 797-802.

[21] Hentschel H.G.E., Deutch J.M. *Phys. Rev.* A, 1984, v. 29, № 3, p. 1609-1611.

[22] Kozlov G.V., Ovcharenko E.N., Mikitaev A.K. *Structure of Polymers Amorphous State. Moscow,* Publishers of the D.I. Mendeleev RKhTU, 2009, 392 p.

[23] Yanovskii Yu.G., Kozlov G.V. Mater. VII Intern. Sci.-Pract. Conf. "New Polymer Composite Materials." *Nal'chik, KBSU,* 2011, p. 189-194.

[24] Wu S. J. *Polymer Sci.: Part B: Polymer Phys.,* 1989, v. 27, № 4, p. 723-741.

[25] Aharoni S.M. *Macromolecules,* 1983, v. 16, № 9, p. 1722-1728.

[26] Budtov V.P. *The Physical Chemistry of Polymer Solutions.* Sankt-Petersburg, Khimiya, 1992, 384 p.

[27] Aharoni S.M. *Macromolecules,* 1985, v. 18, № 12, p. 2624-2630.

[28] Bobryshev A.N., Kozomazov V.N., Babin L.O., Solomatov V.I. *Synergetics of Composite Materials.* Lipetsk, NPO ORIUS, 1994, 154 p.

[29] Kozlov G.V., Yanovskii Yu.G., Karnet Yu. N. *Mekhanika Kompozitsionnykh Materialov i Konstruktsii,* 2005, v. 11, № 3, p. 446-456.

[30] Sheng N., Boyce M.C., Parks D.M., Rutledge G.C., Abes J.I., Cohen R.E. *Polymer,* 2004, v. 45, № 2, p. 487-506.

[31] Witten T.A., Meakin P. *Phys. Rev. B,* 1983, v. 28, № 10, p. 5632-5642.

[32] Witten T.A., Sander L.M. *Phys. Rev. B, 1983,* v. 27, № 9, p. 5686-5697.

[33] Happel J., Brenner G. Hydrodynamics at Small Reynolds Numbers. Moscow, *Mir,* 1976, 418 p.

[34] Mills N.J. *J. Appl. Polymer Sci.,* 1971, v. 15, № 11, p. 2791-2805.

[35] Kozlov G.V., Yanovskii Yu.G., Mikitaev A.K. *Mekhanika Kompozitnykh Materialov, 1998,* v. 34, № 4, p. 539-544.

[36] Balankin A.S. *Synergetics of Deformable Body.* Moscow, Publishers of Ministry Defence SSSR, 1991, 404 p.

[37] Hornbogen E. *Intern. Mater. Res.,* 1989, v. 34, № 6, p. 277-296.

[38] Pfeifer P. *Appl. Surf. Sci.,* 1984, v. 18, № 1, p. 146-164.

[39] Avnir D., Farin D., Pfeifer P. *J. Colloid Interface Sci.,* 1985, v. 103, № 1, p. 112-123.

[40] Ishikawa K. J. *Mater. Sci. Lett.,* 1990, v. 9, № 4, p. 400-402.

[41] Ivanova V.S., Balankin A.S., Bunin I. Zh., Oksogoev A.A. *Synergetics and Fractals in Material Science.* Moscow, Nauka, 1994, 383 p.

[42] Vstovskii G.V., Kolmakov L.G., Terent'ev V.E. *Metally,* 1993, № 4, p. 164-178.

[43] Hansen J.P., Skjeitorp A.T. *Phys. Rev. B,* 1988, v. 38, № 4, p. 2635-2638.

[44] Pfeifer P., Avnir D., Farin D. *J. Stat. Phys., 1984,* v. 36, № 5/6, p. 699-716.

[45] Farin D., Peleg S., Yavin D., Avnir D. Langmuir, 1985, v. 1, № 4, p. 399-407.

[46] Kozlov G.V., Sanditov D.S. Anharmonical Effects and Physical-Mechanical Properties of Polymers. Novosibirsk, *Nauka,* 1994, 261 p.

[47] Bessonov M.I., Rudakov A.P. *Vysokomolek. Soed. B,* 1971, V. 13, № 7, p. 509-511.

[48] Kubat J., Rigdahl M., Welander M. *J. Appl. Polymer Sci.,* 1990, v. 39, № 5, p. 1527-1539.

[49] Yanovskii Yu.G., Kozlov G.V., Kornev Yu.V., Boiko O.V., Karnet Yu.N. *Mekhanika Kompozitsionnykh Materialov i Konstruktsii,* 2010, v. 16, № 3, p. 445-453.

[50] Yanovskii Yu.G., Kozlov G.V., Aloev V.Z. *Mater. Intern. Sci.-Pract. Conf.* "Modern Problems of APK Innovation Development Theory and Practice." Nal'chik, KBSSKhA, 2011, p. 434-437.

[51] Chow T.S. *Polymer,* 1991, v. 32, № 1, p. 29-33.

[52] Ahmed S., Jones F.R. *J. Mater. Sci.,* 1990, v. 25, № 12, p. 4933-4942.

[53] Kozlov G.V., Yanovskii Yu.G., Aloev V.Z. *Mater. Intern. Sci.-Pract. Conf.,* dedicated to FMEP 50-th Anniversary. Nal'chik, KBSSKhA, 2011, p. 83-89.

[54] Andrievskii R.A. *Rossiiskii Khimicheskii Zhurnal,* 2002, v. 46, № 5, p. 50-56.

[55] Kozlov G.V., Sultonov N.Zh., Shoranova L.O., Mikitaev A.K. *Naukoemkie Tekhnologii,* 2011, v. 12, № 3, p. 17-22.

In: Advanced Nanotube and Nanofiber Materials   ISBN: 978-1-62081-170-2
Editors: A. K. Haghi and G. E. Zaikov   © 2012 Nova Science Publishers, Inc.

Chapter 4

CONDUCTIVE CARBON NANOTUBE/NANOFIBER COMPOSITE

A. K. Haghi[*]

University of Guilan, Iran

1. INTRODUCTION

Over the recent decades, scientists became interested in creation of polymer nanofibers due to their potential in many engineering and medical properties [1]. According to various outstanding properties such as very small fiber diameters, large surface area per mass ratio, high porosity along with small pore sizes and flexibility, electrospun nanofiber mats have found numerous applications in diverse areas. For example in the biomedical field, nanofibers play a substantial role in tissue engineering [2], drug delivery [3], and wound dressing [4]. Electrospinning is a novel and efficient method by which fibers with diameters in nanometer scale entitled as nanofibers can be achieved. In electrospinning process, a strong electric field is applied on a droplet of polymer solution (or melt) held by its surface tension at the tip of a syringe needle (or a capillary tube). As a result, the pendent drop will become highly electrified, and the induced charges are distributed over its surface. Increasing the intensity of electric field, the surface of the liquid drop will be distorted to a conical shape known as the Taylor cone [5]. Once the electric

---

[*] E-mail: Haghi@Guilan.ac.ir.

field strength exceeds a threshold value, the repulsive electric force dominates the surface tension of the liquid, and a stable jet emerges from the cone tip. The charged jet is then accelerated toward the target and rapidly thins and dries as a result of elongation and solvent evaporation. As the jet diameter decreases, the surface charge density increases, and the resulting high repulsive forces split the jet to smaller jets. This phenomenon may take place several times leading to many small jets. Ultimately, solidification is carried out, and fibers are deposited on the surface of the collector as a randomly oriented nonwoven mat [6-7]. Figure 1 shows a schematic illustration of electrospinning setup.

The physical characteristics of electrospun nanofibers such as fiber diameter depend on various parameters, which are mainly divided into three categories: solution properties (solution viscosity, solution concentration, polymer molecular weight, and surface tension), processing conditions (applied voltage, volume flow rate, spinning distance, and needle diameter), and ambient conditions (temperature, humidity, and atmosphere pressure) [9]. Numerous applications require nanofibers, with desired properties suggesting the importance of the process control. This end may not be achieved with having a comprehensive outlook of the process and quantitative study of the effects of governing parameters. In this context, Sukigara et al. [10] were assessed the effect of concentration on diameter of electrospun nanofibers. They indicated that the silk nanofibers diameter increases with increasing concentration.

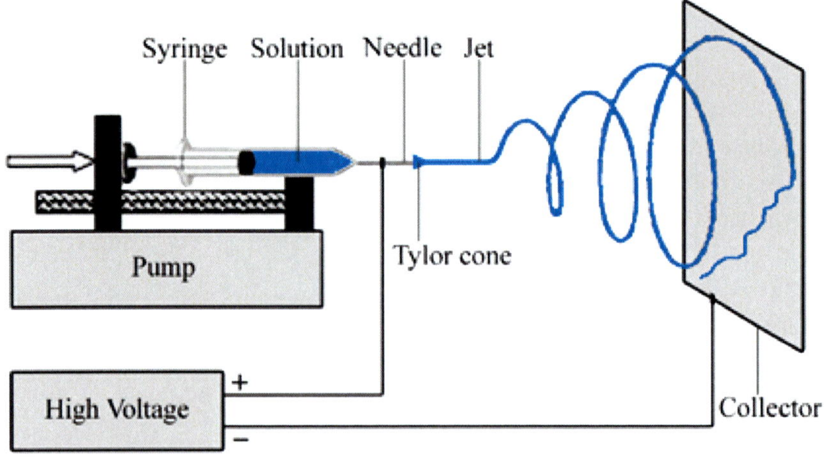

Figure 1. A typical image of Electrospinning process [8].

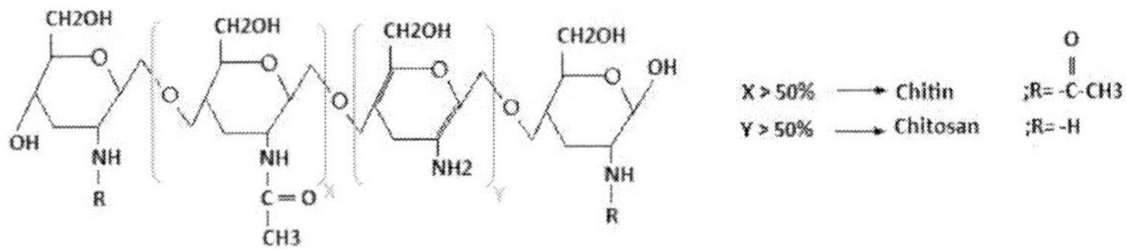

Scheme 1. Chemical structures of Chitin and Chitosan biopolymers.

Besides physical characteristics, medical scientists showed a remarkable attention to biocompatibility and biodegradability of nanofibers made of biopolymers such as collagen [11], fibrogen [12], gelatin [13], silk [14], chitin [15] and chitosan [16]. Chitin is the second abundant natural polymer in the world, and Chitosan (poly-(1-4)-2-amino-2-deoxy-β-D-glucose) is the de-acetylated product of chitin [17]. CHT is well known for its biocompatible and biodegradable properties [18].

Chitosan is insoluble in water, alkali, and most mineral acidic systems. However, though its solubility in inorganic acids is quite limited, chitosan is in fact soluble in organic acids, such as dilute aqueous acetic, formic, and lactic acids. Chitosan also has free amino groups, which make it a positively charged polyelectrolyte. This property makes chitosan solutions highly viscous and complicates its electrospinning [19]. Furthermore, the formation of strong hydrogen bonds in a 3-D network prevents the movement of polymeric chains exposed to the electrical field [20].

Different strategies have been used for bringing chitosan in nanofiber form. The three top most abundant techniques includes blending of favorite polymers for electrospinning process with CHT matrix [21-22], alkali treatment of CHT backbone to improve electrospinnability through reducing viscosity [23] and employment of concentrated organic acid solution to produce nanofibers by decreasing of surface tension [24]. Electrospinning of Polyethylene oxide (PEO)/CHT [21] and polyvinyl alcohol (PVA)/CHT [22] blended nanofiber are two recent studies based on first strategy. In second protocol, the molecular weight of chitosan decreases through alkali treatment. Solutions of the treated chitosan in aqueous 70–90% acetic acid produce nanofibers with appropriate quality and processing stability [23].

Using concentrated organic acids such as acetic acid [24] and tri-flouroacetic acid (TFA) with and without dichloromethane (DCM) [25-26] reported exclusively for producing neat CHT nanofibers. They similarly reported the decreasing of surface tension and at the same time enhancement of charge density of chitosan solution without significant effect on viscosity. This new method suggested significant influence of the concentrated acid solution on the reducing of the applied field required for electrospinning.

The mechanical and electrical properties of neat CHT electrospun natural nanofiber mat can be improved by addition of the synthetic materials including carbon nanotubes (CNTs) [27]. CNTs are one of the important synthetic polymers that were discovered by Iijima in 1991 [28]. CNTs, either single walled nanotubes (SWNTs) or Multi-walled nanotubes (MWNTs), combine together the physical properties of diamond and graphite. They are extremely thermally conductive like diamond and appreciably electrically conductive like graphite. Moreover, the flexibility and exceptional specific surface area to mass ratio can be considered as significant properties of CNTs [29]. The scientists are becoming more interested in CNTs for existence of exclusive properties such as superb conductivity [30] and mechanical strength for various applications. To our knowledge, there has been no report on electrospinning of CHT/MWNTs blend, except for several reports [30-31] that use PVA to improve spinnability. CNTs grafted by CHT were fabricated by electrospinning process. In these novel sheath-core nanofibers, PVA aqueous solution has been used for enhancing nanofiber formation of MWNTs/CHT. Results showed uniform and porous morphology of the electrospun membrane. Despite adequate spinnability, total removing of PVA from nanofiber structure to form conductive substrate is not feasible. Moreover, the structural morphology and mechanical stiffness is extremely affected by thermal or alkali solution treatment of CHT/PVA/MWNTs nanofibers. The chitosan/ carbon nanotube composite can be produced by the hydrogen bonds due to hydrophilic positively charged polycation of chitosan due to amino groups and hydrophobic negatively charged of carbon nanotube due to carboxyl, and hydroxyl groups [32-34].

In the current study, it has been attempted to produce a CHT/MWNTs nanofiber without association of any type of easy electrospinnable polymers. Also, a new approach has been explored to provide highly stable and homogenous composite spinning solution of CHT/MWNTs in concentrated organic acids. This in turn presents a homogenous conductive CHT scaffold, which is extremely important for biomedical implants.

2. EXPERIMENTAL

2.1. Materials

Chitosan polymer with degree of deacetylation of 85% and molecular weight of $5_\times 10^5$ was supplied by Sigma-Aldrich. The MWNTs , supplied by Nutrino, have an average diameter of 4 nm and purity of about 98%. All of the other solvents and chemicals were commercially available and used as received without further purification.

2.2. Preparation of CHT-MWNTs Dispersions

A Branson Sonifier 250 operated at 30W was used to prepare the MWNT dispersions in CHT /organic acid (90% wt acetic acid, 70/30 TFA/DCM) solution based on different protocols. In the first approach (current work) for preparation of sample 1, same amount (3 mg) as received MWNTs were dispersed into deionized water or DCM using solution sonicating for 10 min. Different amount of CHT was then added to MWNTs dispersion for preparation of a 8-12 wt % solution and then sonicated for another 5 min. Figure 2 shows two different protocols used in this study

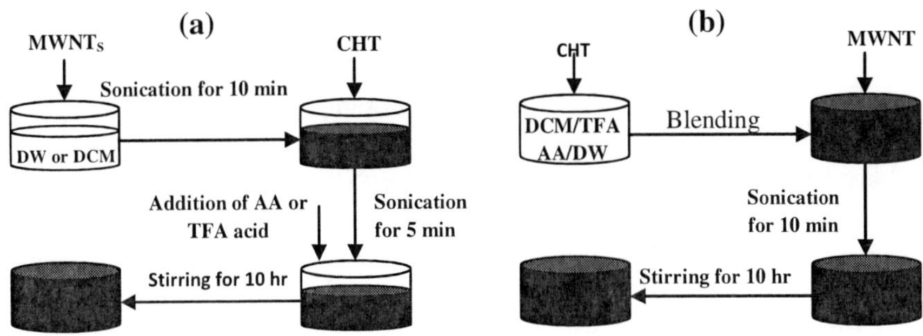

Figure 2. Two protocols used in this study for preparation of MWNTs/CHT dispersion (a) Current study(b) Ref [35].

In the next step, organic acid solution was added to obtain a CHT/MWNT solution with total volume of 5 mL, and finally the dispersion was stirred for another 10 hours. The sample 2 was prepared using second technique. The same amount of MWNTs were dispersed in chitosan solution, and the blend with total volume of 5mL were sonicated for 10 min and dispersion was stirred for 10 hr [35].

2.2. Electrospinning of Chitosan/carbon Nanotube Dispersion

After the preparation of spinning solution, it was transferred to a 5 ml syringe and became ready for spinning of nanofibers. The experiments were carried out on a horizontal electrospinning setup shown schematically in Figure 1. The syringe containing CHT/MWNTS solution was placed on a syringe pump (New Era NE-100) used to dispense the solution at a controlled rate. A high voltage DC power supply (Gamma High Voltage ES-30) was used to generate the electric field needed for electrospinning. The positive electrode of the high voltage supply was attached to the syringe needle via an alligator clip, and the grounding electrode was connected to a flat collector wrapped with aluminum foil where electrospun nanofibers were accumulated to form a nonwoven mat. The voltage and the tip-to-collector distance were fixed respectively on 18-24 kV and 4-10 cm. The electrospinning was carried out at room temperature. Subsequently, the aluminum foil was removed from the collector.

2.3. Measurements and Characterizations

A small piece of mat was placed on the sample holder and gold sputter-coated (Bal-Tec). Thereafter, the micrograph of electrospun PVA fibers was obtained using scanning electron microscope (SEM, Phillips XL-30). Fourier transform infrared spectra (FTIR) were recorded using a Nicolet 560 spectro-meter to investigate the interaction between CHT and MWNT in the range of 800-4000 cm^{-1} under a transmission mode. The size distribution of the dispersed particle was evaluated with a Zetasizer (Malvern Instruments). The conductivity of the conductive fibres was measured using the four point-probe technique. A homemade four-probe electrical conductivity cell operated at constant humidity has been employed. The electrodes were circular pins with separation distance of 0.33 cm, and fibres were connected to pins by silver paint (SPI). Between the two outer electrodes, a constant DC current was applied by Potentiostat/Galvanostat model 363 (Princeton Applied Research). The generated potential difference between the inner electrodes along the current flow direction was recorded by digital multi-meter 34401A (Agilent).

The conductivity (δ: S/cm) of the nanofiber thin film with rectangular surface can then be calculated according to length (L:cm), width (W:cm), thickness (t:cm), DC current applied (mA) and the potential drop across the two inner electrodes (mV). All measuring repeated at least five times for each set of samples.

$$\delta = \frac{I \times L}{V \times W \times t} \tag{1}$$

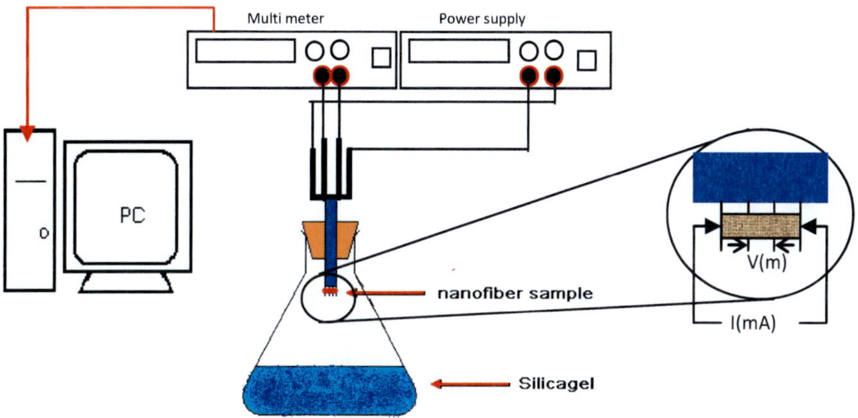

Figure 3. The experimental setup for four-probe electrical conductivity measurement of nanofiber thin film.

3. RESULTS AND DISCUSSION

3.1. The Characteristics of MWNT/CHT Dispersion

Utilisation of MWNTs in biopolymer matrix initially requires their homogenous dispersion in a solvent or polymer matrix. Dynamic light scattering (DLS) is a sophisticated technique used for evaluation of particle size distribution. DLS provides many advantages as a particle size analysis method that measures a large population of particles in a very short time period, with no manipulation of the surrounding medium. Dynamic light scattering of MWNTs dispersions indicates that the hydrodynamic diameter of the nanotube bundles is between 150 and 400 nm after 10 min of sonication for sample 2. (Figure 4)

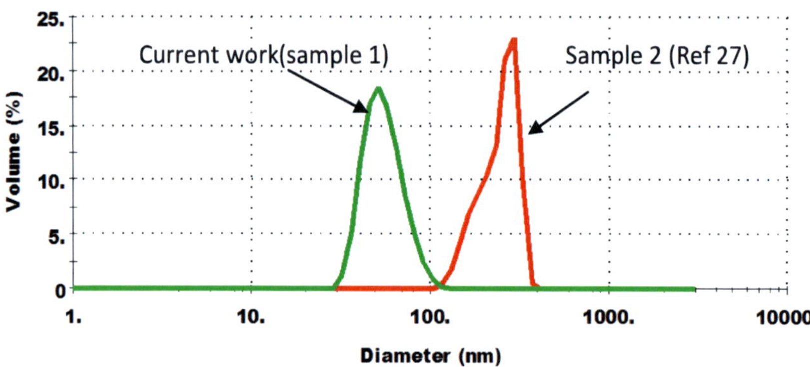

Figure 4. Hydrodynamic diameter distribution of MWNT bundles in CHT/acetic acid (1%) solution for different preparation technique.

Figure 5. Stability of CHT-MWNT dispersions (a) Current work (sample1)(b) Ref [35].

MWNTs bundle in sample 1(different approach but same sonication time compared to sample 2) shows a range of hydrodynamic diameter between 20-100 nm. (Figure 4). The lower range of hydrodynamic diameter for sample 1 can be correlated to more exfoliated and highly stable nanotubes strands in

CHT solution. The higher stability of sample 1 compared to sample 2 over a long period of time is confirmed by solution stability test. The results presented in Figure 3 indicate that procedure employed for preparation of sample 1 (current work) was an effective method for dispersing MWNTs in CHT/acetic acid solution. However, MWNTs bundles in sample 2 were found to re-agglomerate upon standing after sonication, as shown in Figure 5, where the sedimentation of large agglomerated particles is indicated.

Despite the method reported in ref 27, neither sedimentation nor aggregation of the MWNTs bundles was observed in first sample. Presumably, this behavior in sample 1 can be attributed to contribution of CHT biopolymer to form an effective barrier against reagglomeration of MWNTs nanoparticles. In fact, using sonication energy, in first step without presence of solvent, makes very tiny exfoliated but unstable particle in water as dispersant. Instantaneous addition of acetic acid as solvent to prepared dispersion and long mixing most likely helps the wrapping of MWNTs strands with CHT polymer chain.

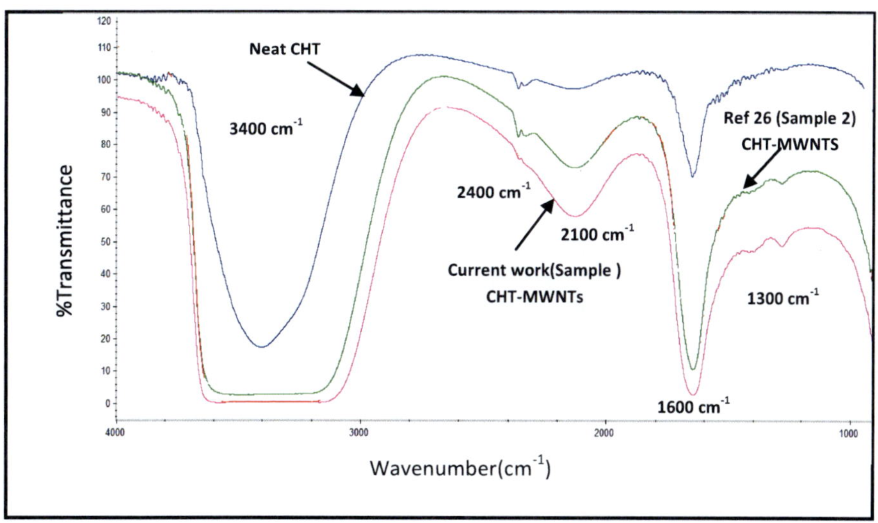

Figure 6. FTIR spectra of CHT-MWCNT in 1% acetic acid with different techniques of dispersion.

Figure 6 shows the FTIR spectra of neat CHT solution and CHT/MWNTs dispersions prepared using strategies explained in experimental part. The interaction between MWNTs and CHT in dispersed form has been understood through recognition of functional groups. The enhanced peaks at ~1600 cm^{-1}

can be attributed to (N-H) band and (C=O) band of amid functional group. However, the intensity of amid group for CHT/MWNTs dispersion has been increased, presumably due to contribution of G band in MWNTs. More interestingly, in this region, the FTIR spectra of MWNTs-CHT dispersion (sample 1) have been highly intensified compared to sample 2 [35]. It can be correlated to higher chemical interaction between acid functionalized C-C group of MWNTs and amid functional group in CHT.

This probably is the main reason of the higher stability and lower MWNTs dimension demonstrated in Figure 4 and Figure 5. Moreover, the intensity of protonated secondary amine absorbance at 2400 cm^{-1} for sample prepared by new technique is negligible compared to sample 2 and neat CHT. Furthermore, the peak at 2123cm^{-1} is a characteristic band of the primary amine salt, which is associated with the interaction between positively charged hydrogen of acetic acid and amino residues of CHT. Also, the broad peaks at ~3410 cm^{-1} due to the stretching vibration of OH group superimposed on NH stretching bond and broaden due to inter hydrogen bonds of polysaccharides. The broadest peak of hydrogen bonds was observed at 3137-3588 cm^{-1} for MWNTs/CHT dispersion prepared by new technique (sample 1).

3.2. The Physical and Morphological Characteristics of MWNTs/CHT Nanofiber

The different solvents including acetic acid 1-90%, pure formic acid, and TFA/DCM tested for the electrospinning of chitosan/carbon nanotube. No jet was seen upon applying the high voltage even above 25 kV by using of acetic acid 1-30% and formic acid as the solvent for chitosan/carbon nanotube. When the acetic acid 30-90% was used as the solvent, beads were deposited on the collector. Therefore, under these conditions, an electrospun fiber of carbon nanotube/chitosan could not be obtained (data not shown).

Figure 7 shows scanning electronic micrographs of the MWNTs/CHT electrospun nanofibers in different concentration of CHT in TFA/DCM (70:30) solvent. As presented in Figure 7a, at low concentrations of CHT, the beads deposited on the collector and thin fibers coexited among the beads. When the concentration of CHT was increased as shown in Figures 7a-c, the beads were decreased. Figure 7c show homogenous electrospun nanofibers with minimum beads, thin fibers and interconnected fibers. More increasing of concentration of CHT lead to increasing of interconnected fibers at Figures 7 d-e. Figure 8 show the effect of concentration on average diameter of MWNTs/CHT electrospun nanofibers. Our assessments indicate that the fiber diameter of MWNTs/CHT increases with the increasing concentration. In this context, there are several studies that have reported results similar to our results [36-37]. Hence, MWNTs/CHT solution in TFA/DCM (70:30) with 10 wt% chitosan resulted as optimization conditions of concentration for electrospinning. An average diameter of 275 nm (Figure 7c: diameter distribution, 148-385) investigated for this conditions. Table 1 list the variation of nanofiber diameter and four-probe electrical conductivity based on the different loading of CHT. One can expect the lower conductivity once the CHT content increases. However, this effect has been damped by decreasing of nanofiber diameter. This led to a nearly constant conductivity over entire measurements.

To understand the effects of voltage on morphologies of CHT/MWNT electrospun nanofibers, the SEM images at Figure 9 were analyzed. In our experiments, 18 kv were attained as threshold voltage, where fiber formation occurred. When the voltage was low, the beads and some little fiber deposited on collector (Figure 9a). As shown in Figures 9a-d, the beads decreased by increasing voltage from 18 kV to 24 kV for electrospinning of fibers. The nanofibers collected by applying 18 kV (9a) and 20 kV (9b) were not quite clear and uniform. The higher the applied voltage, the more uniform nanofibers with less distribution starts to form. The average diameter of fibers, 22 kV (9c), and 24 kV (9d), respectively, were 204 (79-391), and 275 (148-385).

Table 1. The variation of conductivity and mean nanofiber diameter versus Chitosan loading

% CHT (%w/v)	% MWNT (%w/v)	Voltage (KV)	Tip to collector (cm)	Diameter (nm)	Conductivity (S/cm)
8	0.06	24	5	137 ± 58	NA
9	0.06	24	5	244 ± 61	9×10^{-5}
10	0.06	24	5	275 ± 70	9×10^{-5}
11	0.06	24	5	290 ± 87	8×10^{-5}
12	0.06	24	5	Non uniform	NA

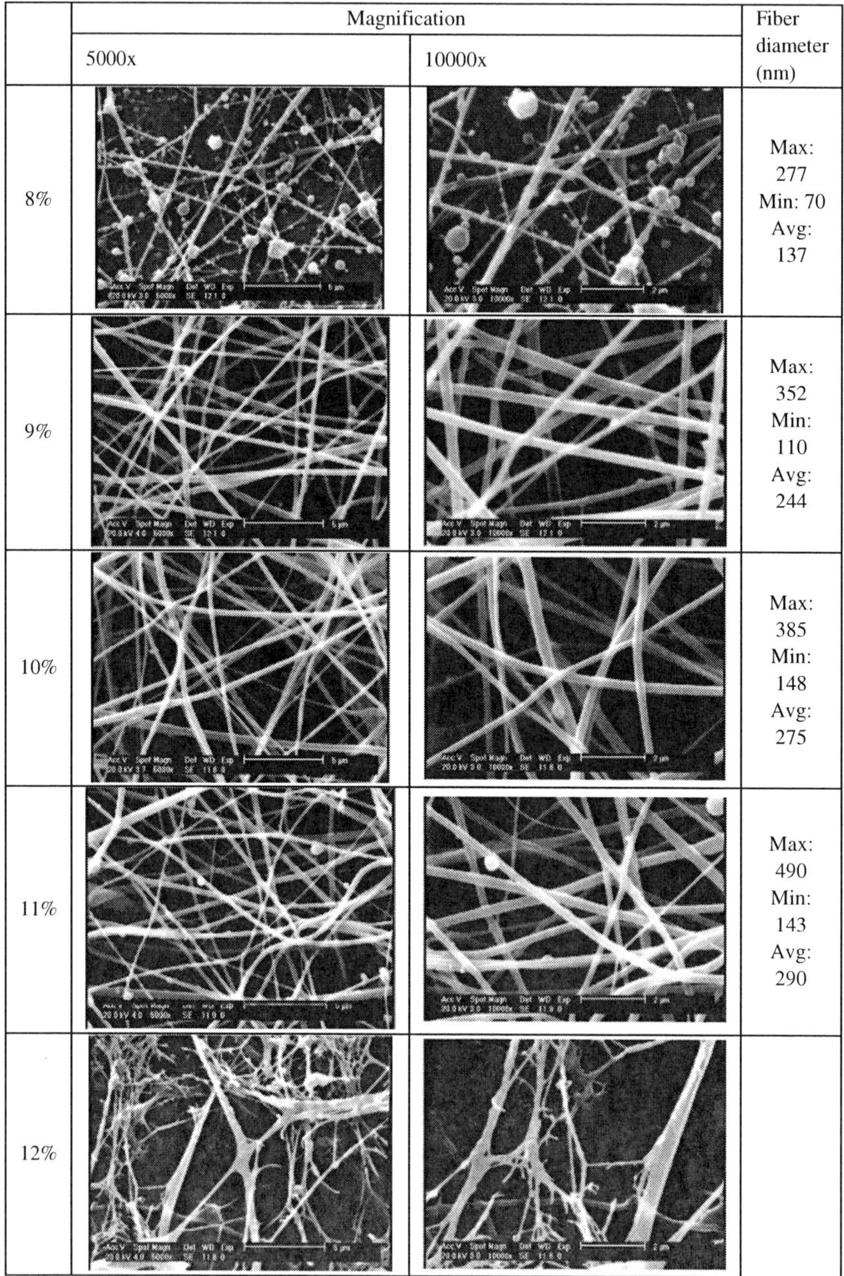

	Magnification		Fiber diameter (nm)
	5000x	10000x	
8%			Max: 277 Min: 70 Avg: 137
9%			Max: 352 Min: 110 Avg: 244
10%			Max: 385 Min: 148 Avg: 275
11%			Max: 490 Min: 143 Avg: 290
12%			

Figure 7. Scanning electron micrographs of electrospun nanofibers at different CHT concentration (wt%): (a) 8, (b) 9, (c) 10, (d) 11, (e) 12, 24 kV, 5 cm, TFA/DCM: 70/30, (0.06%wt MWNTs).

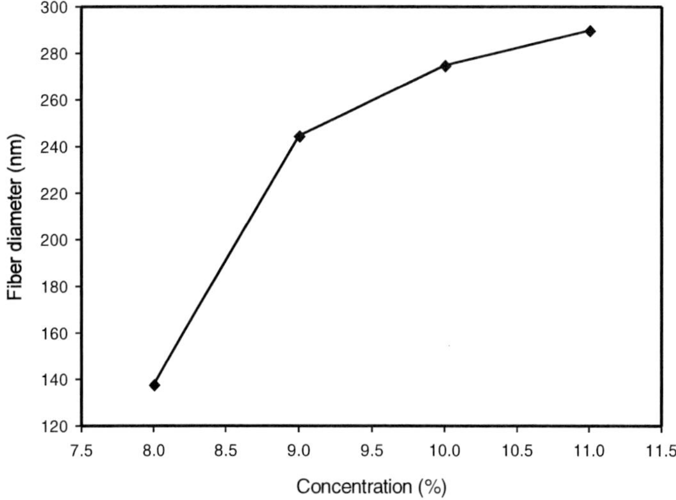

Figure 8. The effect of the concentration of CHIT/CNT dispersion on fiber diameter.

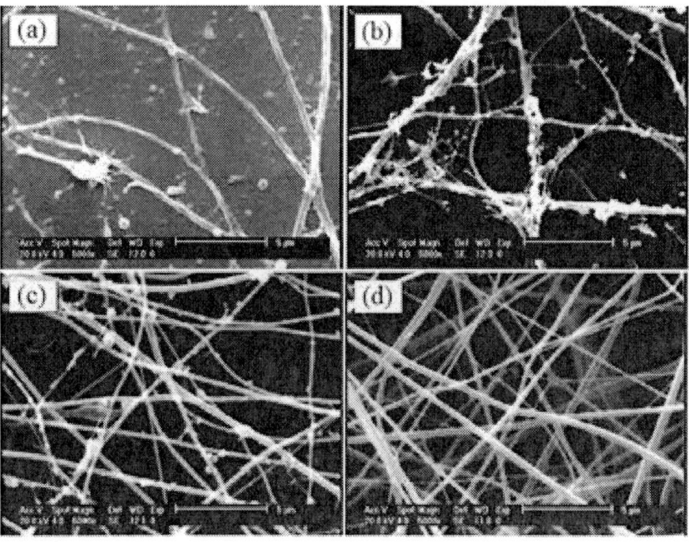

Figure 9. Scanning electronic micrographs of electrospun fibers at different voltage (kV): (a) 18, (b) 20, (c) 22, (d) 24, 5 cm, 10 wt%, TFA/DCM: 70/30. (0.06%wt MWNTs).

**Table 2. The variation of conductivity and mean nanofiber
diameter versus applied voltage**

% CHT (%w/v)	% MWNT (%w/v)	Voltage (KV)	Tip to collector (cm)	Diameter (nm)	Conductivity (S/cm)
10	0.06	18	5	Non uniform	NA
10	0.06	20	5	Non uniform	NA
10	0.06	22	5	201 ± 66	6×10^{-5}
10	0.06	24	5	275 ± 70	9×10^{-5}

The conductivity measurement given in Table 2 confirms our observation in first set of conductivity data. As can be seen from last row, the amount of electrical conductivity reaches to a maximum level of 9×10^{-5} at prescribed setup.

The distance between the tip to collector is another approach that controls the fiber diameter and morphology. Figure 10 shows the change in morphologies of CHT/MWNTs electrospun nanofibers at different distances between the tip to collector. When the distance tip-to-collector is not long enough, the solvent could not be vaporized, hence, a little interconnected thick fiber deposits on the collector (Figure 10a). In the 5 cm distance of tip-to-collector (Figure 10b), rather homogenous nanofibers have obtained with negligible beads and interconnected fibers. However, the beads increased by increasing of distance tip-to-collector as represented from Figure 10b to Figure 10f. Similar results were observed for chitosan nanofibers fabricated by Geng et al. [24]. Also, the results show that the diameter of electrospun fibers decreased by increasing of distance tip to collector in Figures 10b, 10c, 10d, respectively, 275 (148-385), 170 (98-283), 132 (71-224). Similar effect of distance between tip-to-collector on fiber diameter has observed in previous studies [38-39]. A remarkable defects and nonhomogenity appears for those fibers prepared at a distance of 8 cm (Figure 10e) and 10 cm (Figure 10f). However, 5 cm for distance tip-to-collector was seen as proper for electrospinning.

Conductivity results also are in agreement with those data obtained in previous parts. The nonhomogenity and huge bead densities play as a barrier against electrical current, and still a bead-free and thin nanofiber mat shows higher conductivity compared to other samples. Experimental framework in this study was based on parameter adjusting for electrospinning of conductive CHT/MWNTs nanofiber. It can be expected that the addition of nanotubes can boost conductivity and also change morphological aspects, which is extremely important for biomedical applications.

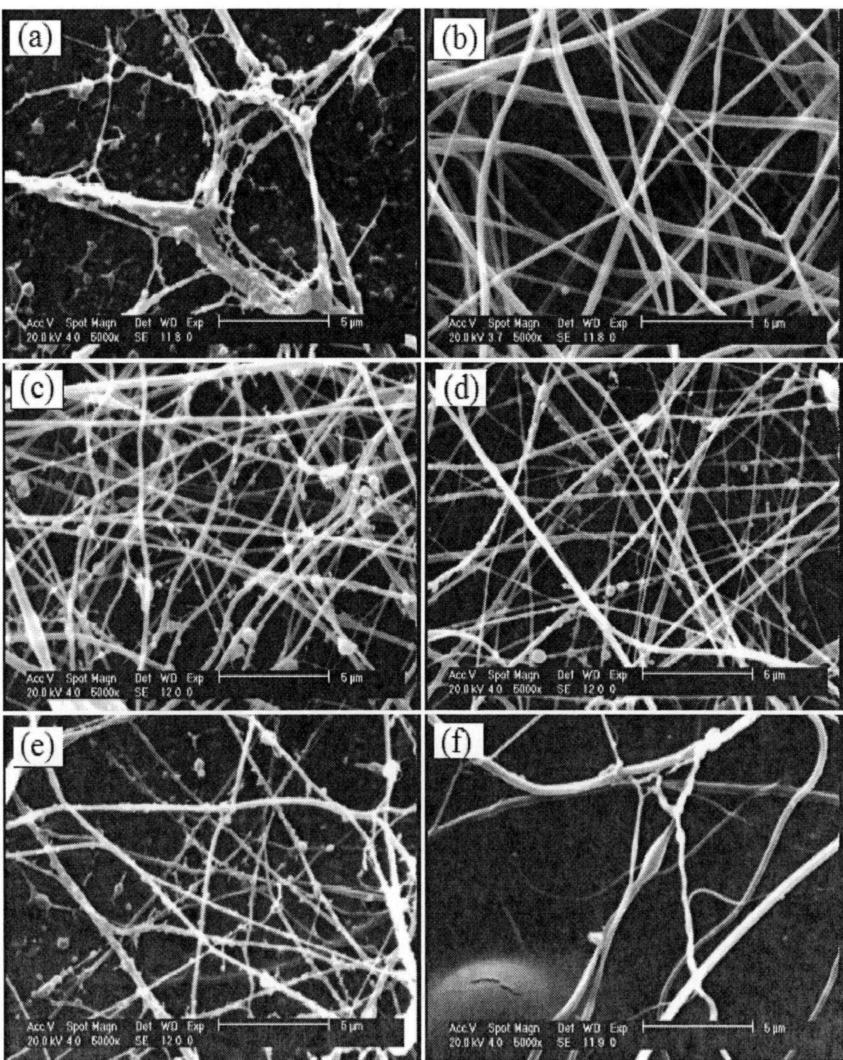

Figure 10. Scanning electronic micrographs of electrospun fibers of Chitosan/Carbon nanotubes at different tip-to-collector distances (cm): (a) 4, (b) 5, (c) 6, (d) 7, (e) 8, (f) 10, 24 kV, 10 wt%, TFA/DCM: 70/30.

Table 3. The variation of conductivity and mean nanofiber diameter versus applied voltage

% CHT (%w/v)	% MWNT (%w/v)	Voltage (KV)	Tip to collector (cm)	Diameter (nm)	Conductivity (S/cm)
10	0.06	24	4	Non uniform	NA
10	0.06	24	5	275 ± 70	9×10^{-5}
10	0.06	24	6	170 ± 58	6×10^{-5}
10	0.06	24	7	132 ± 53	7×10^{-5}
10	0.06	24	8	Non uniform	NA
10	0.06	24	10	Non uniform	NA

CONCLUSION

Conductive composite nanofiber of CHT/MWNTs has been produced using conventional electrospinning technique. A new protocol is suggested for preparation of electrospinning solution, which shows much better stability and homogeneity compared previous techniques. Several solvents including acetic acid 1-90%, formic acid, and TFA/DCM (70:30) were investigated in the electrospinning of CHT/MWNTs dispersion. It was observed that the TFA/DCM (70:30) solvent is most preferred for fiber formation process with acceptable electrospinnability. The formation of nanofibers with conductive pathways regarding to exfoliated and interconnected nanotube strands is a breakthrough in chitosan nanocomposite area. This can be considered as a significant improvement in electrospinning of chitosan/carbon nanotube dispersion. It has been also observed that the homogenous fibers with an average diameter of 275 nm could be prepared with a conductivity of 9×10^{-5}.

REFERENCES

[1]   Agarwal, S., J.H. Wendorff, and A. Greiner, *Use of electrospinning technique for biomedical applications.* Polymer, 2008. 49(26): p. 5603-5621.

[2]   Li, M. et al., *Electrospun protein fibers as matrices for tissue engineering.* Biomaterials, 2005. 26(30): p. 5999-6008.

[3]   Zeng, J. et al., *Influence of the drug compatibility with polymer solution on the release kinetics of electrospun fiber formulation.* Journal of Controlled Release, 2005. 105(1-2): p. 43-51.

[4]   Khil, M.-S. et al., *Electrospun nanofibrous polyurethane membrane as wound dressing.* Journal of Biomedical Materials Research Part B: Applied Biomaterials, 2003. 67B(2): p. 675-679.

[5]   G.I.Taylor, *Electrically driven jets.* Proc Roy Soc London 1969. 313: p. 453-475.

[6]   Doshi, J. and D.H. Reneker, *Electrospinning process and applications of electrospun fibers.* Journal of Electrostatics, 1995. 35(2-3): p. 151-160.

[7]   Li, D. and Y. Xia, *Electrospinning of Nanofibers: Reinventing the Wheel?* Advanced Materials, 2004. 16(14): p. 1151-1170.

[8]   Ziabari, M., V. Mottaghitalab, and A.K. Haghi, *Evaluation of electrospun nanofiber pore structure parameters.* Korean J. Chem. Eng., 2008. 25(4): p. 923-932.

[9]   Tan, S.H. et al., *Systematic parameter study for ultra-fine fiber fabrication via electrospinning process.* Polymer, 2005. 46(16): p. 6128-6134.

[10]   Sukigara, S. et al., *Regeneration of Bombyx mori silk by electrospinning—part 1: processing parameters and geometric properties.* Polymer, 2003. 44: p. 5721-5727.

[11]   Matthews, J.A. et al., *Electrospinning of collagen nanofibers.* Biomacromolecules, 2002. 3(2): p. 232-8.

[12]   McManus, M.C. et al., *Electrospun fibrinogen: Feasibility as a tissue engineering scaffold in a rat cell culture model.* Journal of Biomedical Materials Research Part A, 2007. 81A(2): p. 299-309.

[13]   Huang, Z.-M. et al., *Electrospinning and mechanical characterization of gelatin nanofibers.* Polymer, 2004. 45(15): p. 5361-5368.

[14]   Zhang, X., M.R. Reagan, and D.L. Kaplan, *Electrospun silk biomaterial scaffolds for regenerative medicine.* Advanced Drug Delivery Reviews, 2009. 61(12): p. 988-1006.

[15]   Noh, H.K. et al., *Electrospinning of chitin nanofibers: Degradation behavior and cellular response to normal human keratinocytes and fibroblasts.* Biomaterials, 2006. 27(21): p. 3934-3944.

[16]   K. Ohkawa et al., *Chitosan Nanofiber.* Biomacromolecules, 2006. 7: p. 3291-3294.

[17]   Agboh, O.C. and Y. Qin, *Chitin and Chitosan Fibers.* Polymers for Advanced Technologies, 1997. 8: p. 355-365.

[18] Rinaudo, M., *Chitin and chitosan: Properties and applications.* Prog. Polym. Sci., 2006. 31: p. 603-632.

[19] Aranaz, I. et al., *Functional Characterization of Chitin and Chitosan.* Current Chemical Biology, 2009. 3: p. 203-230.

[20] Neamnark, A., R. Rujiravanit, and P. Supaphol, *Electrospinning of hexanoyl chitosan.* Carbohydrate Polymers, 2006. 66(3): p. 298-305.

[21] Duan, B. et al., *Electrospinning of chitosan solutions in acetic acid with poly(ethylene oxide).* Vol. 15. 2004, Leiden, PAYS-BAS: Brill. 15.

[22] Jia, Y.T. et al., *Fabrication and characterization of poly (vinyl alcohol)/chitosan blend nanoWbers produced by electrospinning method.* Carbohydrate Polymers, 2007. 67: p. 403-409.

[23] Homayoni, H., S.A.H. Ravandi, and M. Valizadeh, *Electrospinning of chitosan nanofibers: Processing optimization.* Carbohydrate Polymers, 2009. 77: p. 656-661.

[24] Geng, X., O.H. Kwon, and J. Jang, *Electrospinning of chitosan dissolved in concentrated acetic acid solution.* Biomaterials, 2005. 26: p. 5427-5432.

[25] TORRES-GINER et al., *Development of Active Antimicrobial Fiber Based Chitosan Polysaccharide Nanostructures using Electrospinning.* 2008, Weinheim, ALLEMAGNE: Wiley. 12.

[26] Vrieze, S.D. et al., *Electrospinning of chitosan nanofibrous structures: feasibility study.* Journal of Materials Science, 2007. 42(19): p. 8029-8034.

[27] Ohkawa, K. et al., *Electrospinning of Chitosan.* Macromolecular Rapid Communications, 2004. 25(18): p. 1600-1605.

[28] Iijima, S., *Helical microtubules of graphitic carbon.* Nature, 1991. 354(6348): p. 56–58.

[29] Esawi, A.M.K. and M.M. Farag, *Carbon nanotube reinforced composites: Potential and current challenges.* Materials & Design, 2007. 28(9): p. 2394-2401.

[30] Feng, W. et al., *The fabrication and electrochemical properties of electrospun nanofibers of a multi-walled carbon nanotube grafted by chitosan.* Nanotechnology, 2008. 19(10): p. 105707.

[31] Liao, H. et al., *Improved cellular response on multi-walled carbon nanotube-incorporated electrospun polyvinyl alcohol/chitosan nanofibrous scaffolds.* Colloids and Surfaces B: Biointerfaces, 2011. In Press, Accepted Manuscript(doi:10.1016/j.colsurfb.2011.02.010): p. 1-28.

[32] Baek, S.-H., B. Kim, and K.-D. Suh, *Chitosan particle/multi-wall carbon nanotube composites by electrostatic interactions.* Colloids and Surfaces A: Physicochemical and Engineering Aspects, 2008. 316(1-3): p. 292-296.

[33] Liu, Y.-L., W.-H. Chen, and Y.-H. Chang, *Preparation and properties of chitosan/carbon nanotube nanocomposites using poly(styrene sulfonic acid)-modified CNTs.* Carbohydrate Polymers, 2009. 76(2): p. 232-238.

[34] Tkac, J., J.W. Whittaker, and T. Ruzgas, *The use of single walled carbon nanotubes dispersed in a chitosan matrix for preparation of a galactose biosensor.* Biosensors and Bioelectronics, 2007. 22(8): p. 1820-1824.

[35] Spinks, G.M. et al., *Mechanical properties of chitosan/CNT microfibers obtained with improved dispersion.* Sensors and Actuators B: Chemical, 2006. 115(2): p. 678-684.

[36] Zhang, H. et al., *Regenerated-Cellulose/Multi-walled-Carbon-Nanotube Composite Fibers with Enhanced Mechanical Properties Prepared with the Ionic Liquid 1-Allyl-3-methylimidazolium Chloride.* Adv. Mater. , 2007. 19: p. 698–704.

[37] Deitzel, J.M. et al., *The effect of processing variables on the morphology of electrospun nanofibers and textiles.* Polymer, 2001. 42(1): p. 261-272.

[38] Zhang, S., W.S. Shim, and J. Kim, *Design of ultra-fine nonwovens via electrospinning of Nylon 6: Spinning parameters and filtration efficiency.* Materials & Design, 2009. 30(9): p. 3659-3666.

[39] Li, Y., Z. Huang, and Yandong, *Electrospinning of nylon-6,66,1010 terpolymer.* European Polymer Journal, 2006. 42(7): p. 1696-1704.

In: Advanced Nanotube and Nanofiber Materials   ISBN: 978-1-62081-170-2
Editors: A. K. Haghi and G. E. Zaikov   © 2012 Nova Science Publishers, Inc.

Chapter 5

NANOSTRUCTURED FABRICS BASED ON ELECTROSPUN NANOFIBERS

*A. K. Haghi**
University of Guilan, Iran

INTRODUCTION

Nowadays, there are different types of protective clothing—some of are disposable and some are non-disposable. The simplest and most preliminary of this equipment is made from rubber or plastic that is completely impervious to hazardous substances. Unfortunately, these materials are also impervious to air and water vapor and thus retain body heat, exposing their wearer to heat stress, which can build quite rapidly to a dangerous level. Another approach to protective clothing is incorporating activated carbon into multi-layer fabric in order to absorb toxic vapors from environment and prevent penetration to the skin. The use of activated carbon is considered only a short-term solution because it loses its effectiveness upon exposure to sweat and moisture. The use of semi-permeable membranes as a constituent of the protective material is a another approach. In this way, reactive chemical decontaminants encapsulates in microparticles or fills in Microporous Hollow fibers and then coats onto fabric. The microparticle or fiber walls are permeable to toxic vapors but

---

* E-mail: Haghi@Guilan.ac.ir.

impermeable to decontaminants, so that the toxic agents diffuse selectively into them and neutralize [1-3].

Generally, a negative relationship always exists between thermal comfort and protection performance for currently available protective clothing. Thus there still exists a very real demand for improved protective clothing that can offer acceptable levels of impermeability to highly toxic pollutions of low molecular weight, while minimizing wearer discomfort and heat stress.

Electrospinning provides an ultrathin membrane-like web of extremely fine fibers with very small pore size and high porosity, which makes them excellent candidates for use in filtration, membrane, and possibly protective clothing applications. Preliminary investigations have indicated that the using of nanofiber web in protective clothing structure could present minimal impedance to air permeability and extremely efficiency in trapping dust and aerosol particles. Meanwhile, it is found that the electrospun webs of nylon 6,6, polybenzimidazole, polyacrylonitrile, and polyurethane provided good aerosol particle protection, without a considerable change in moisture vapor transport or breathability of the system. While nanofiber webs suggest exciting characteristics, it has been reported that they have limited mechanical properties. In order to provide suitable mechanical properties for use as cloth, nanofiber webs must be laminated via an adhesive into a fabric system. This system could protect ultrathin nanofiber web versus mechanical stresses over an extended period of time [4-14].

The adhesives could be as melt adhesive or solvent-based adhesive. When a melt adhesive is used, the hot-press laminating is carried out at temperatures above the softening or melting point of adhesive. If a solvent-based adhesive is used, laminating process could perform at room temperature. In addition, the solvent-based adhesive is generally environmentally unfriendly, more expensive and usually flammable, whereas the hot-melt adhesive is environmentally friendly, inexpensive requires less heat, and so is now more preferred. However, without disclosure of laminating details, the hot-press method is more suitable for nanofiber web lamination. In this method, laminating temperature is one of the most important parameters. Incorrect selection of this parameter may lead to change or damage nanofiber web. Thus, it is necessary to find out a laminating temperature that has the least effect on the nanofiber web.

It has been found that morphology such as fiber diameter and its uniformity of the electrospun polymer fibers are dependent on many processing parameters. These parameters can be divided into three groups as shown in Table 1. Under certain conditions, not only uniform fibers but also beads-like formed fibers can be produced by electrospinning. Although the parameters of the electrospinning process have been well analyzed in each of polymers, this information has been inadequate enough to support the electrospinning of ultra-fine nanometer scale polymer fibers. A more systematic parametric study is hence required to investigate.

Table 1. Processing parameters in electrospinning

Solution properties	Viscosity
	Polymer concentration
	Molecular weight of polymer
	Electrical conductivity
	Elasticity
	Surface tension
Processing conditions	Applied voltage
	Distance from needle to collector
	Volume feed rate
	Needle diameter
Ambient conditions	Temperature
	Humidity
	Atmospheric pressure

The purpose of this study is to consider the influence of laminating temperature on nanofiber/laminate properties. Multi-layer fabrics were made by electrospinning polyacrylonitrile nanofibers onto nonwoven substrate and incorporating into fabric system via hot-press method at different temperatures.

EXPERIMENTAL

Electrospinning and Laminating Process

Polyacrylonitrile (PAN) of 70,000 g/mol molecular weight from Polyacryl Co. (Isfehan, Iran) has been used with Dimethylformamide (DMF) from Merck to form a polymer solution 12% w/w after stirring for 5h and exposing for 24h at ambient temperature. The yellow and ripened solution was inserted into a plastic syringe with a stainless steel nozzle 0.4 mm in inner diameter, and then it was placed in a metering pump from *World Precision Instruments* (Florida, USA). Next, this set was installed on a plate, which it could traverse to left-right along drum (Figure 1). The flow rate 1 µl/h for solution was selected, and the fibers were collected on an aluminum-covered rotating drum (with speed 9 m/min), which was previously covered with a polypropylene spun-bond nonwoven (PPSN) substrate of 28cm× 28cm dimensions; 0.19 mm thickness; 25 g/m2 weight; 824 $cm^3/s/cm^2$ air permeability and 140°C melting point. The distance between the nozzle and the drum was 7cm, and an electric voltage of approximately 11kV was applied between them. Electrospinning process was carried out for 8h at room temperature to reach approximately web thickness 3.82 g/m². Then nanofiber webs were laminated into cotton weft-warp fabric with a thickness 0.24mm and density of 25×25 (warp-weft) per centimeter to form a multi-layer fabric (Figure 2). Laminating was performed at temperatures 85,110,120,140,150°C for 1 min under a pressure of 9 gf/cm^2.

Nanofiber Web Morphology

In order to consider nanofiber web morphology after hot-pressing, another laminating was performed by a non-stick sheet made of Teflon (0.25 mm thickness) instead one of the fabrics (fabric /pp web/nanofiber web/pp web/non-stick sheet). Finally, after removing of Teflon sheet, the nanofiber layer side was observed under an optical microscope (MICROPHOT-FXA, Nikon, Japan) connected to a digital camera.

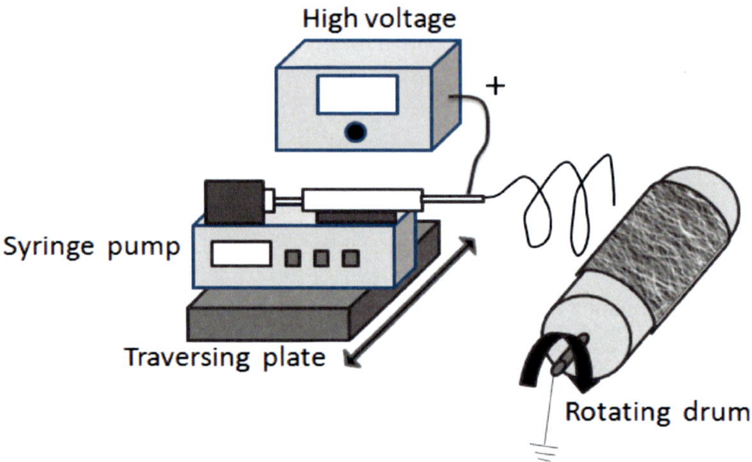

Figure 1. Electrospinning setup.

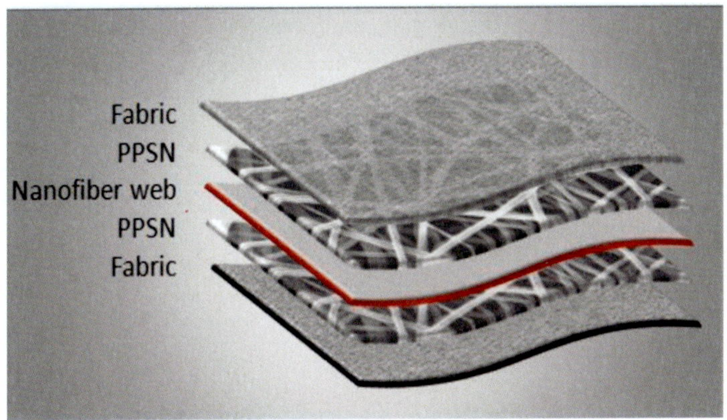

Figure 2. Multi-layer fabric components.

Measurement of Air permeability

Air permeability of multi-layer fabric after lamination was tested by TEXTEST FX3300 instrument (Zürich, Switzerland). It was tested for five pieces of each sample under air pressure 125pa at ambient condition (16°C, 70%RH) and obtained average air permeability.

Mechanical Properties of Multi-layer Nanoweb

The tensile strength of multi-layer fabrics with and without nanofiber web were carried out using MICRO250 tensile machine (SDL International Ltd.). Ten samples were cut from the warp directions of multi-layer fabric at size of 10mm×200mm and then exposed to the standard condition (25°C,60% RH) for 24h in order to condition. To measure tensile strength, testing was performed by load cell of 25Kgf. Also, the distance between the jaws and the rate of extension were selected 100mm and 20mm/min, respectively.

RESULTS AND DISCUSSION

PPSN was selected as melt adhesive layer for hot-press laminating (Figure 2). This process was performed under different temperatures to find an optimum condition.

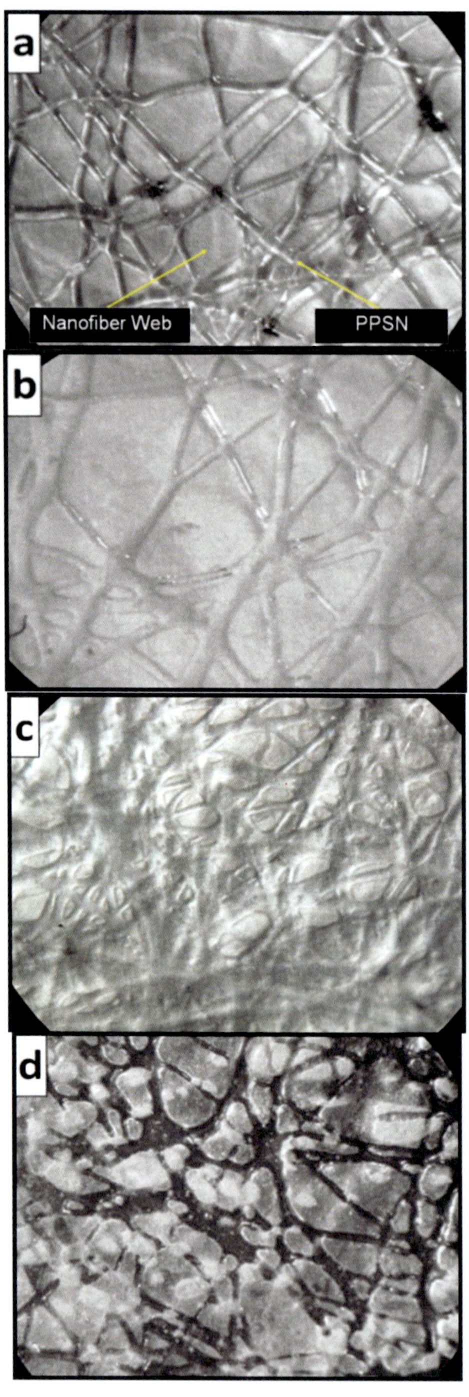
a
Nanofiber Web
PPSN
b
c
d

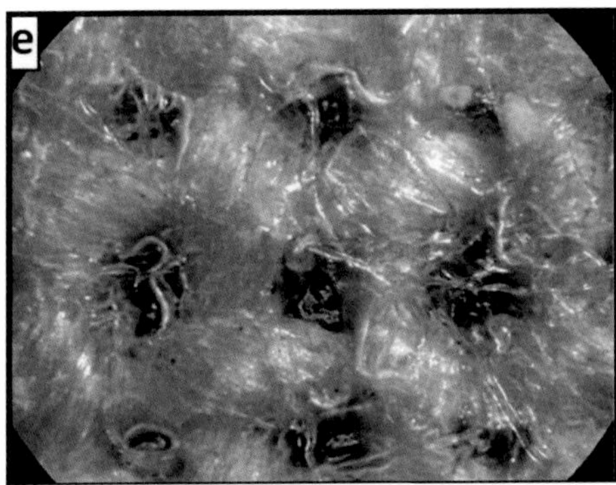

Figure 3. a) The optical microscope images of nanofiber web after laminating at 85°C (at 100 magnification); b) The optical microscope images of nanofiber web after laminating at 110°C (at 100 magnification); c) The optical microscope images of nanofiber web after laminating at 120°C (at 100 magnification); d) The optical microscope images of nanofiber web after laminating at 140°C (at 100 magnification); e) The optical microscope images of nanofiber web after laminating at temperatures more than 140°C (at 100 magnification)

Figure 3 presents the optical microscope images of nanofiber web after lamination. It is obvious that by increasing laminating temperature to melting point (samples a-c), the adhesive layer gradually melts and spreads on web surface. But when melting point selected as laminating temperature (sample d), the nanofiber web begin to damage. In this case, the adhesive layer completely melted and penetrated into nanofiber web and occupied its pores. This procedure intensified by increasing of laminating temperature above melting point. As shown in Figure 1 (sample e), perfect absorption of adhesive by nanofiber web creates a transparent film, which leads to appear fabric structure.

Breathability

Also, to examine how laminating temperature affects the breathability of multi-layer fabric, air permeability experiment was performed. Figure 4 indicates the effect of laminating temperature on air permeability. As might be expected, air permeability decreased with increasing laminating temperature. This behavior is attributed to melting procedure of adhesive layer. As mentioned above, before melting point, the adhesive gradually spreads on web surface. This phenomenon causes the adhesive layer to act like an impervious barrier to air flow and reduces air permeability of multi-layer fabric. But at melting point and above, the penetration of melt adhesive into nanofiber/fabric structure leads to fill its pores and finally decrease in air permeability.

Adhesion of Layers

Furthermore, we only observed that the adhesive force between layers was increased according to the temperature rise. The sample (a) exhibited very poor adhesion between nanofiber web and fabric, and it could be separated by light abrasion of thumb, while adhesion increased by increasing laminating temperature to melting point. It must be noted that after melting point, because of passing of melt PPSN across nanofiber web, adhesion between two layers of fabric will occur.

Tensile Strength

The tensile strength of samples without nanofibers (Figure 5) is weaker than those laminated with nanofibers (Figure 6). According to Table 2, the breaking load and breaking elongation for the samples laminated with electrospun nanofibers are improved as well. These variations can be observed clearly in Figures 7 and 8 for ten samples.

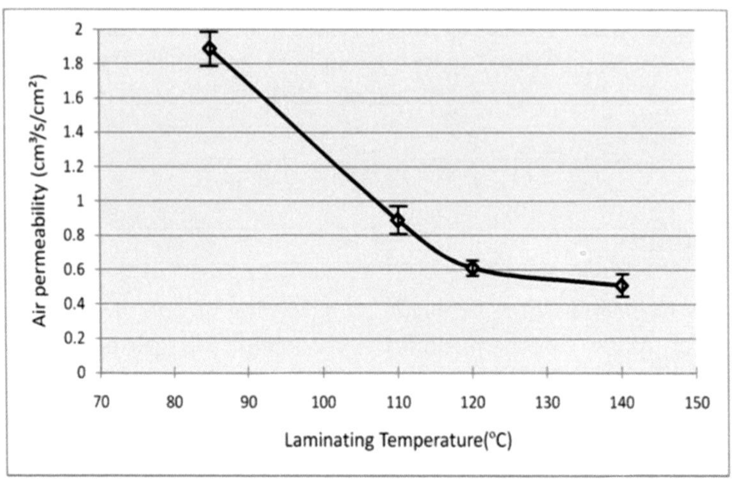

Figure 4 . Air permeability of multi-layer fabric as a function of laminating temperature.

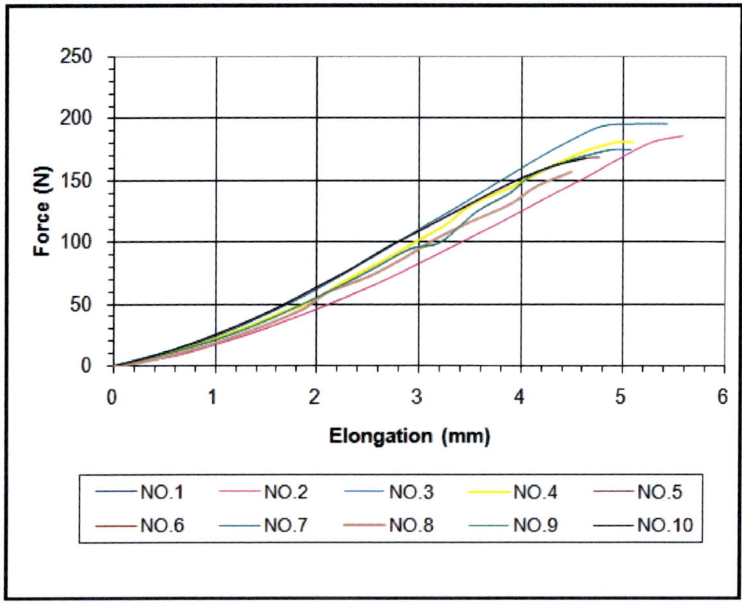

Figure 5. Force-Elongation curve for multi-layer fabric without Nanofiber web.

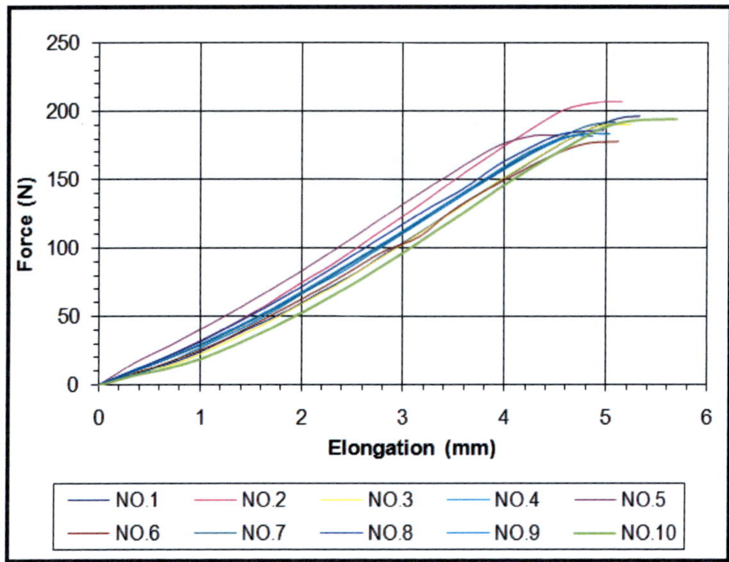

Figure 6. Force-Elongation curve for multi-layer fabric with Nanofiber web.

Table 2. Tensile strength test results of the Multi-layer fabrics

Multi-layer Fabric	Warp direction			
	Breaking Load, N		Breaking Elongation, mm	
	Mean value	CV, %	Mean value	CV, %
Without Nanofiber web	174.427	6.2	5.02	7.5
With nanofiber web	189.211	4.6	5.11	6

Simulation of Nanoweb

For the continuous fibers, it is assumed that the lines are infinitely long so that in the image plane, all lines intersect the boundaries. Under this scheme (Figure), a line with a specified thickness is defined by the perpendicular distance d from a fixed reference point O located in the center of the image, and the angular position of the perpendicular α. Distance d is limited to the diagonal of the image.

Based on the objective of this chapter, several variables are allowed to be controlled during the simulation:

1. *Web density* that can be controlled using the line density, which is the number of lines to be generated in the image.
2. *Angular density*, which is useful for generating fibrous structures with specific orientation distribution. The orientation may be sampled from either a normal or a uniform random distribution.
3. *Distance from the reference point* normally varies between zero and the diagonal of the image, restricted by the boundary of the image and is sampled from a uniform random distribution.
4. *Line thickness* (fiber diameter) is sampled from a normal distribution. The mean diameter and its standard deviation are needed.
5. *Image size* can also be chosen as required.

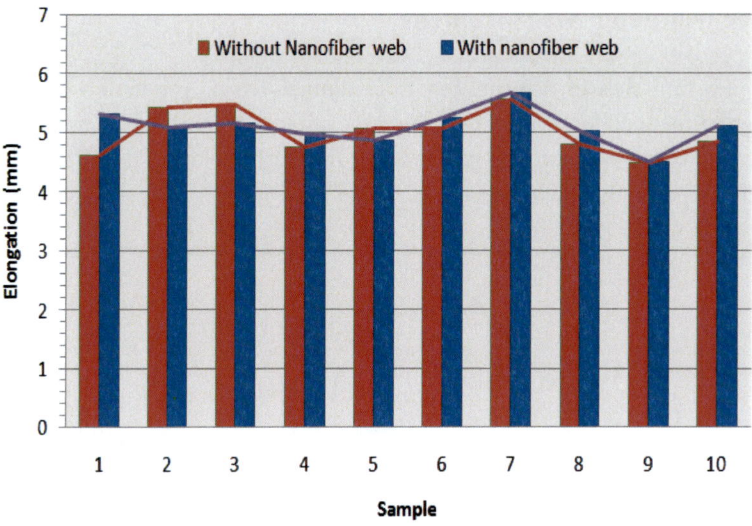

Figure 7. Breaking elongation of ten samples.

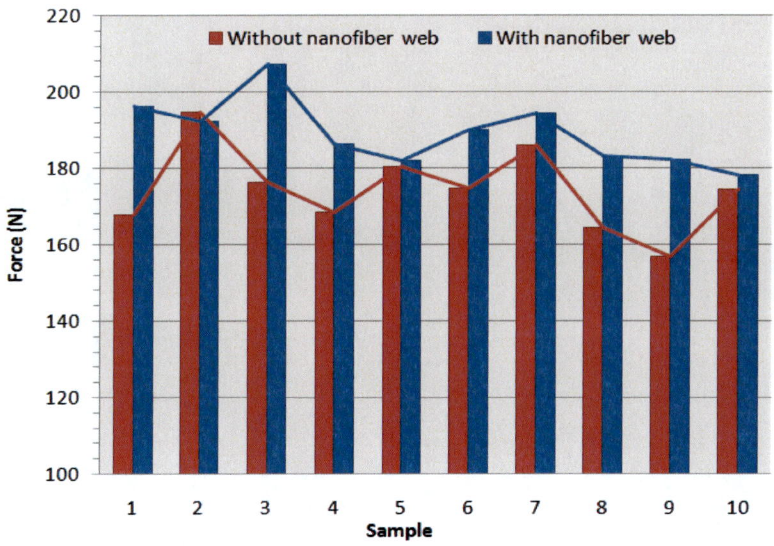

Figure 8. Breaking load of ten samples.

Fiber Diameter Measurement

The first step in determining fiber diameter is to produce a high-quality image of the web, called micrograph, at a suitable magnification using electron microscopy techniques. The methods for measuring electrospun fiber diameter are described in following sections.

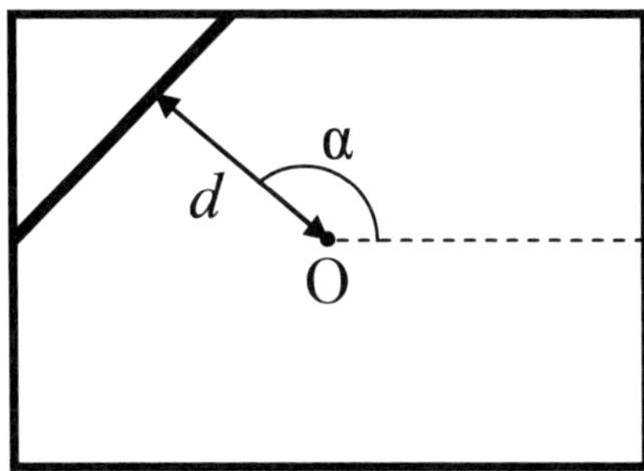

Figure 9. Procedure for μ-randomness.

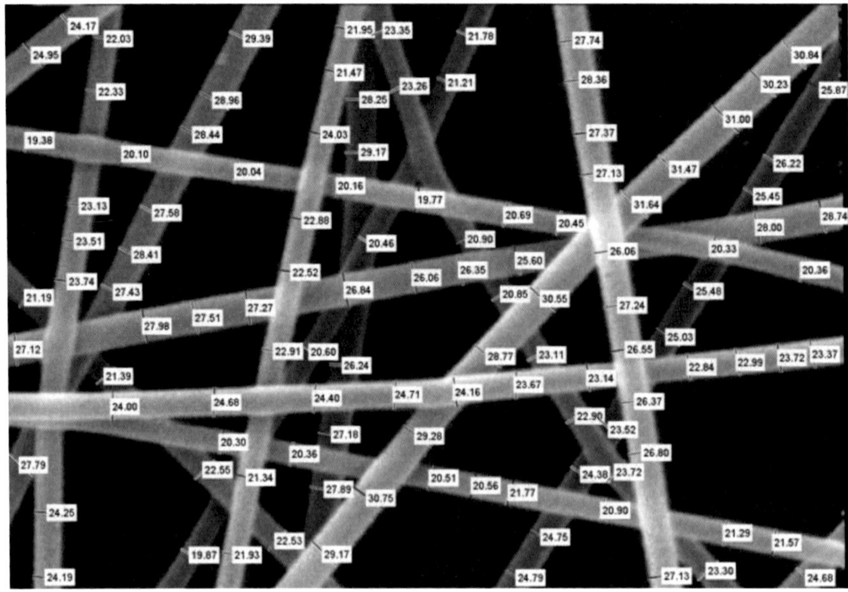

Figure 10. Manual method.

Manual Method

The conventional method of measuring the fiber diameter of electrospun webs is to analyze the micrograph manually. The manual analysis usually consists of determining the length of a pixel of the image (setting the scale), identifying the edges of the fibers in the image and counting the number of pixels between two edges of the fiber (the measurements are made perpendicular to the direction of fiber-axis), converting the number of pixels to *nm* using the scale and recording the result. Typically, 100 measurements are carried out (Figure 10).

However, this process is tedious and time consuming, especially for large number of samples. Furthermore, it cannot be used as on-line method for quality control since an operator is needed for performing the measurements. Thus, developing automated techniques that eliminate the use of operator and have the capability of being employed as on-line quality control is of great importance.

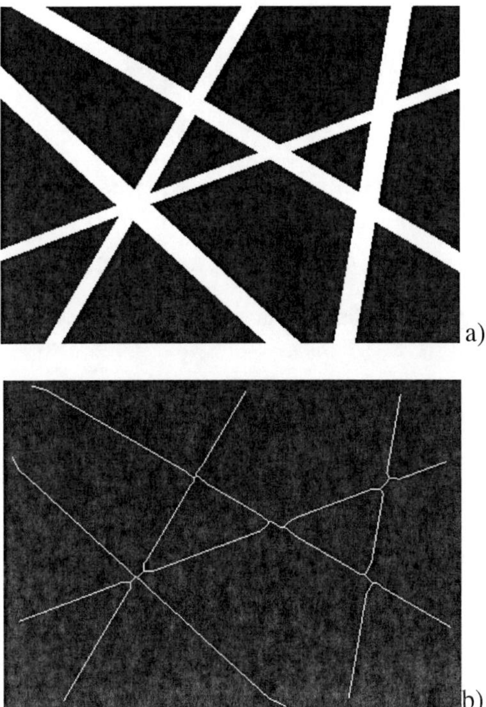

Fig. 11 (Continued)

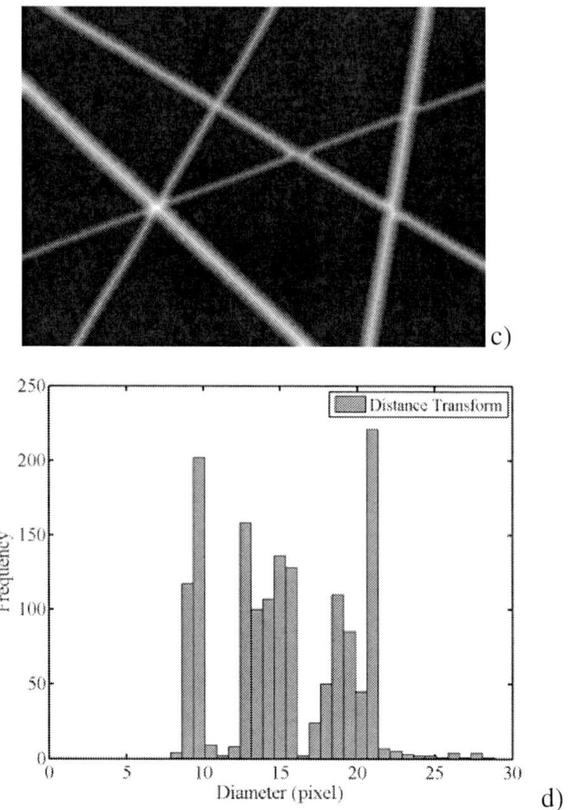

Figure 11. a) A simple simulated image, b) Skeleton of (a), c) Distance map of (a) after pruning, d) Histogram of fiber diameter distribution obtained by distance transform method.

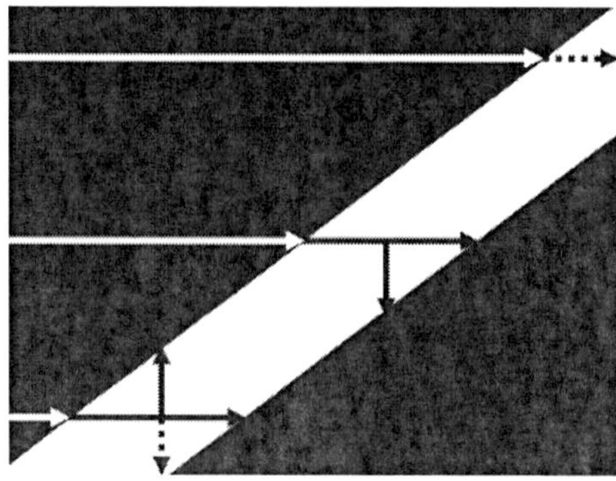

Figure 12. Diameter measurement based on two scans in direct tracking method.

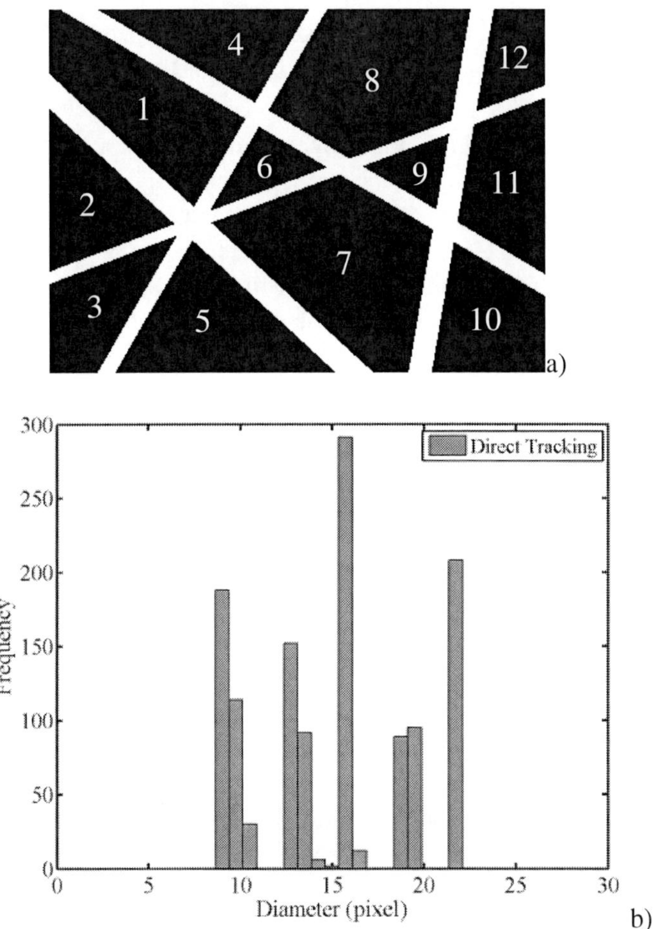

Figure 13. a) A simple simulated image, which is labeled, b) Histogram of fiber diameter distribution obtained by direct tracking.

Distance Transform

The *distance transform* of a binary image is the distance from every pixel to the nearest nonzero-valued pixel. The center of an object in the distance transformed image will have the highest value and lie exactly over the object's *skeleton*. The skeleton of the object can be obtained by the process of *skeletonization or thinning.* The algorithm removes pixels on the boundaries of objects but does not allow objects to break apart. This reduces a thick object to its corresponding object with one pixel width. Skeletonization or thinning often produces short spurs, which can be cleaned up automatically with a *pruning* procedure.

The algorithm for determining fiber diameter uses a binary input image and creates its skeleton and distance transformed image. The skeleton acts as a guide for tracking the distance transformed image by recording the intensities to compute the diameter at all points along the skeleton. Figure 11 shows a simple simulated image, which consists of five fibers with diameters of 10, 13, 16, 19 and 21 pixels, together with its skeleton and distance map including the histogram of fiber diameter obtained by this method.

Direct Tracking

Direct tracking method uses a binary image as an input data to determine fiber diameter based on information acquired from two scans—first a horizontal and then a vertical scan. In the horizontal scan, the algorithm searches for the first white pixel adjacent to a black. Pixels are counted until reaching the first black. The second scan is then started from the midpoint of horizontal scan, and pixels are counted until the first black is encountered. Direction changes if the black pixel isn't found. Having the number of horizontal and vertical scans, the number of pixels in perpendicular direction, which is the fiber diameter, could be measured from a geometrical relationship. The explained process is illustrated in Figure 12.

In electrospun nonwoven webs, nanofibers cross each other at intersection points, and this brings about the possibility for some untrue measurements of fiber diameter in these regions. To circumvent this problem, a process called *fiber identification* is employed. First, black regions are labeled and couple of regions between which a fiber exists is selected. In the next step, the two selected regions are connected performing a *dilation* operation with a large enough *structuring element*. Dilation is an operation that grows or thickens objects in a binary image by adding pixels to the boundaries of objects. The specific manner and extent of this thickening is controlled by the size and shape of the structuring element used. In the following process, an *erosion* operation with the same structuring element is performed, and the fiber is re-cognized. Erosion shrinks or thins objects in a binary image by removing pixels on object boundaries. As in dilation, the manner and extent of shrinking is controlled by a structuring element. Then, in order to enhance the processing speed, the image is cropped to the size of selected regions. Afterwards, fiber diameter is measured according to the previously explained algorithm. This trend is continued until all of the fibers are analyzed. Finally, the data in pixels may be converted to *nm*, and the histogram of fiber diameter distribution is plotted. Figure 13 shows a labeled simple simulated image and the histogram of fiber diameter obtained by this method.

CONCLUSION

In this chapter, effect of laminating temperature on nanofiber/laminate properties were investigated to make next-generation protective clothing. First, surface images of nanofiber web after lamination were taken using optical microscope in order to consider morphology changes. It was observed that nanofiber web remains unchanged as laminating temperature is below PPSN melting point. In addition, to compare breathability of laminates, air permeability was measured. It was found that by increasing laminating temp, air permeability was decreased. Furthermore, it only was observed that the adhesive force between layers in laminate was increased with temperature rise. The mechanical properties of the samples laminated by electrospun nanofibers showed significant improvements.

These results indicated that laminating temperature is effective parameter for lamination of nanofiber web into fabric structure. Thus, by varying this parameter, fabrics with different levels of thermal comfort and protection could be developed depending on our need and use.

REFERENCES

[1]   M. Ziabari, V. Mottaghitalab, S. T. McGovern and A. K. Haghi, *Chim. Phys. Lett.*, 25, 3071 (2008).

[2]   M. Ziabari, V. Mottaghitalab, S. T. McGovern and A. K. Haghi, *Nanoscale Research Letter*, 2, 297(2007).

[3]   M. Ziabari, V. Mottaghitalab and A. K. Haghi, *Korean J. Chem. Eng.*, 25, 919 (2008).

[4]   M. Ziabari, V. Mottaghitalab and A. K. Haghi, *Korean J. Chem. Eng.*, 25, 923 (2008).

[5]   M. Ziabari, V. Mottaghitalab and A. K. Haghi, *Korean J. Chem. Eng.*, 25, 905 (2008).

[6]   A. K. Haghi and M. Akbari, *Physica Status Solidi*, 204, 1830 (2007).

[7]   M. Kanafchian, M. Valizadeh and A.K. Haghi, *Korean J. Chem. Eng.*, 28, 428 (2011).

[8]   M. Kanafchian, M. Valizadeh and A.K. Haghi, *Korean J. Chem. Eng.*, 28, 763 (2011).

[9]   M. Kanafchian, M. Valizadeh and A.K. Haghi, *Korean J. Chem. Eng.*, 28, 751 (2011).

[10]   M. Kanafchian, M. Valizadeh and A.K. Haghi, *Korean J. Chem. Eng.*, 28, 445(2011).

[11]   A. Afzali, V. Mottaghitalab, M. Motlagh, A.K. Haghi, *Korean J. Chem. Eng.*, 27, 1145(2010).

[12]   Z. Moridi, V. Mottaghitalab, A.K. Haghi, *Korean J. Chem. Eng.*, 28, 445(2011).

[13]   A.K. Haghi, *Cellulose Chem. Technol.*, 44, 343 (2010)

[14]   Z. Moridi, V. Mottaghitalab, A. K. Haghi, *Cellulose Chem. Technol.*, 45, 549 (2011)

In: Advanced Nanotube and Nanofiber Materials   ISBN: 978-1-62081-170-2
Editors: A. K. Haghi and G. E. Zaikov   © 2012 Nova Science Publishers, Inc.

Chapter 6

CARBON NANOTUBES GEOMETRY AND REINFORCEMENT DEGREE OF POLYMER NANOCOMPOSITES

Z. M. Zhirikova[1], V. Z. Aloev[1], G. V. Kozlov[1] and G. E. Zaikov[2]

[1]Kabardino-Balkarian State Agricultural Academy,
Nal'chik Russian Federation
[2]N.M. Emanuel Institute of Biochemical Physics
of Russian Academy of Sciences,
Moscow, Russian Federation

INTRODUCTION

At present, it is considered that carbon nanotubes (CNT) are one of the most perspective nanofillers for polymer nanocomposites [1]. The high anisotropy degree (their length to diameter large ratio) and low transverse stiffness are CNT_s-specific features. These factors define CNT_s ring-like structures formation at manufacture and their introduction in polymer matrix. Such structures radius depends to a considerable extent on CNT_s length and diameter. Thus, the strong dependence of nanofiller structure on its geometry is CNT_s application-specific feature. Therefore, the present work purpose is to study the dependence of nanocomposites butadiene-styrene rubber/carbon nanorubes (BSR/CNT) properties on nanofiller structure, received by CVD method with two catalysts usage.

EXPERIMENTAL

The nanocomposites BSR/CNT with CNT content of 0.3 mass % have been used as the study object. CNT have been received in the Institute of Applied Mechanics of Russian Academy of Sciences by the vapors catalytic chemical deposition method (CVD), based on carbon-containing gas thermochemical deposition on non metallic catalyst surface. Two catalysts – Fe/Al_2O_3 (CNT-Fe) and Co/Al_2O_3 (CNT-Co) – have been used for the studied CNT. The received nanotubes have diameter of 20 nm and length of order of 2 mcm.

The nanofiller structure was studied on force-atomic microscope Nano-DST (Pacific Nanotechnology, USA) by a semi-contact method in the force modulation regime. The received CNT size and polydispersity analysis was made with the aid of the analytical disk centrifuge (CPS Instrument, Inc., USA), allowing to determine with high precision the size and distribution by sizes in range from 2 nm up to 5 mcm. The nanocomposites BSR/CNT elasticity modulus was determined by nanoindentation method on apparatus Nano-Test 600 (Great Britain).

RESULTS AND DISCUSSION

In Figure 1, the electron microphotographs of CNT coils are adduced, which demonstrate ring-like structures formation for this nanofiller. In Figure 2, the indicated structures distribution by sizes was shown, from which it follows that for CNT-Fe, narrow enough monodisperse distribution with maximum at 280 nm is observed and for CNT-Co – polydisperse distribution with maximums at ~ 50 and 210 nm.

Further, let us carry out the analytical estimation of CNT formed ring-like structures radius R_{CNT}. The first method uses the following formula, obtained within the frameworks of percolation theory [2]:

$$\varphi_n = \frac{\pi L_{\text{CNT}} r_{\text{CNT}}^2}{\left(2R_{\text{CNT}}\right)^3},$$ (1)

where φ_n is CNT volume content, L_{CNT} and r_{CNT} are CNT length and radius, respectively.

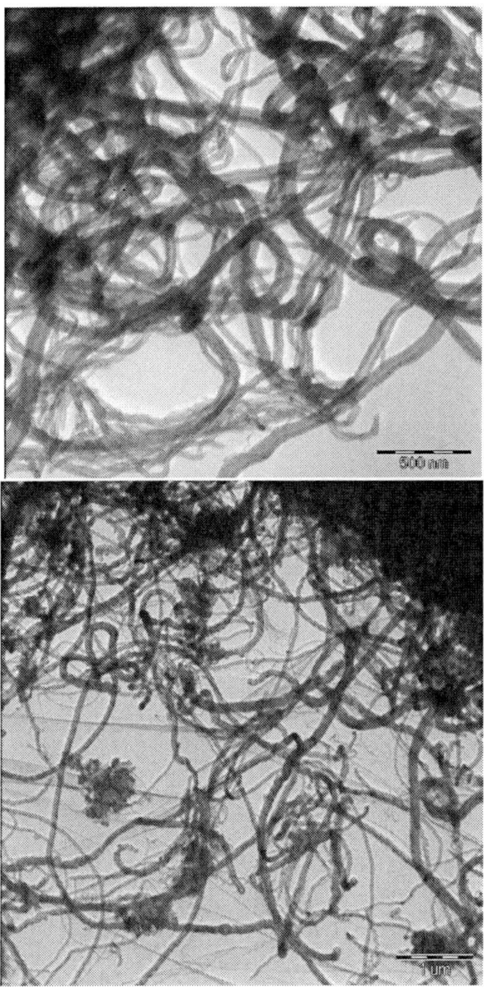

Figure 1. Electron micrographs of CNT structure, received on transmission electron microscope.

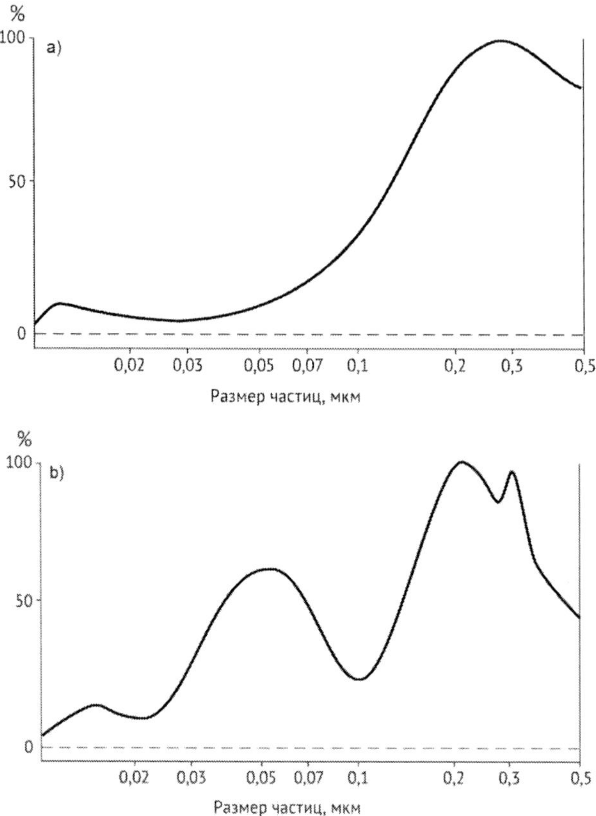

Figure 2. The particles distribution by sizes for CNT-Fe (1) and CNT-Co (2).

The value φ_n was determined according to the well-known equation [3]:

$$\varphi_n = \frac{W_n}{\rho_n},\tag{2}$$

where W_n and ρ_n are mass content and density of nanofiller, respectively.

In its turn, the value ρ_n was calculated as follows [4]:

$$\rho_n = 0.188 D_{CNT}^{1/3},\tag{3}$$

where D_{CNT} is CNT diameter.

The second method is based on the following empirical formula application [5]:

$$R_{CNT} = \left(\frac{D_{CNT}}{D_{CNT}^{st}}\right)^2 \left(0.64 + 4.5 \times 10^{-3} \varphi_n^{-1}\right), \text{ mcm,} \qquad (4)$$

where D_{CNT}^{st} is a standard nanotube diameter, accepted in paper [5] equal to CNT of the mark "Taunite" diameter (45 nm).

The values R_{CNT}, calculated according to the equations (1) and (4), are adduced in Table 1, from which their good correspondence (the discrepancy is equal to ~ 15 %) follows. Besides, they correspond well enough to Figure 2 data.

**Table 1. The structural and mechanical characteristics
of nanocomposites BSR/CNT**

Catalyst	E_n, MPa	E_n/E_m	E_n/E_m, the equation (5)	R_{CNT}, nm, the equation (1)	R_{CNT}, nm, the equation (4)	b, the equation (6)
Fe/Al$_2$O$_3$	4.9	1.485	1.488	236	278	8.42
Co/Al$_2$O$_3$	3.1	~ 1.0	1.002	236	278	0.27

Footnote: the value E_m for BSR is equal to 3.3 MPa.

In Table 1, the values of elasticity modulus E_n for the studied nanocomposites and E_m for the initial BSR are also adduced. As one can see, if for the nanocomposite BSR/CNT-Fe, the very high (with accounting of the condition $W_n=0.3$ mass %) reinforcement degree $E_n/E_m=1.485$ was obtained, then for the nanocomposite BSR/CNT-Co, reinforcement is practically absent (with accounting for experiment error): $E_n \approx E_m$. Let us consider the reasons of such essential distinction.

As it is known [4], the reinforcement degree for nanocomposites polymer/CNT can be calculated as follows:

$$\frac{E_n}{E_m} = 1 + 11\left(c\varphi_n b\right)^{1.7}, \qquad (5)$$

where c is proportionality coefficient between nanofiller φ_n and interfacial regions φ_{if} relative fractions, b is the parameter, characterizing interfacial adhesion polymer matrix-nanofiller level.

The parameter b in the nanocomposites polymer/CNT case depends on nanofiller geometry as follows [5]:

$$b = 80\left(\frac{R_{CNT}^2}{L_{CNT}D_{CNT}^2}\right).$$

(6)

Calculated according to the equation (6), values b (for BSR/CNT-Fe and BSR/CNT-Co R_{CNT} magnitudes were accepted equal to 280 and 50 nm, respectively) are adduced in Table 1. As one can see, R_{CNT} decreasing for the second from the indicated nanocomposites results to b reduction more than 30 times.

The coefficient c value in the equation (5) can be calculated as follows [4]. First, the interfacial layer thickness l_{if} is determined according to the equation [6]:

$$l_{if} = a\left(\frac{r_{CNT}}{a}\right)^{2\left(d - d_{surf}\right)/d},$$

(7)

where a is lower linear scale of polymer matrix fractal behaviour, accepted equal to statistical segment length l_{st}, d is dimension of Euclidean space, in which fractal is considered (it is obvious that in our case $d=3$), d_{surf} is CNT surface dimension, which for the studied CNT was determined experimentally and equal to 2.89.

The indicated dimension d_{surf} has a very large absolute magnitude ($2 \leq d_{surf} < 3$ [4]) that supposes corresponding roughness of CNT surface, which BSR macromolecule, simulated by rigid statistical segments sequence [7], cannot be "reproduced." Therefore in practice, the effective value d_{surf} (d_{surf}^{ef} for $d_{surf} > 2.5$) is used, which is equal to [7]:

$$d_{surf}^{ef} = 5 - d_{surf}.$$

(8)

And at last, the statistical segment length l_{st} is estimated according to the equation [4]:

$$l_{st} = l_0 C_\infty,\tag{9}$$

where l_0 is the length of the main chain skeletal bond, C_∞ is characteristic ratio. For BSR l_0=0.154 nm, C_∞=12.8 [8].

Further, simulating an interfacial layer as cylindrical one with external radius $r_{CNT}+l_{if}$ and internal radius r_{CNT}, let us obtain from geometrical considerations the formula for φ_{if} calculation [6]:

$$\varphi_{if} = \varphi_n \left[\left(\frac{r_{CNT} + l_{if}}{r_{CNT}} \right)^3 - 1 \right],\tag{10}$$

according to which the value c is equal to 3.47.

The reinforcement degree E_n/E_m calculation results according to the equation (5) are adduced in Table 1. As one can see, these results are very close to the indicated parameter experimental estimations. From the equation (5), it follows unequivocally that the values E_n/E_m distinction for nanocomposites BSR/CNT-Fe, and BSR/CNT-Co is defined by the interfacial adhesion level difference only, characterized by the parameter b, since the values c and φ_n are the same for the indicated nanocomposites. In its turn, from the equation (6), it follows so unequivocally that the parameters b distinction for the indicated nanocomposites is defined by R_{CNT} difference only, since the values L_{CNT} and D_{CNT} for them are the same. Thus, the fulfilled analysis supposes CNT geometry plays crucial role in nanocomposites polymer/CNT mechanical properties determination.

Let us note that the usage of the average value R_{CNT} for nanocomposites BSR/CNT-Co according to Figure 2 data in the equation (6), which is equal to 130 nm, will not change the conclusions made above. In this case, E_n/E_m=1.036, which is again close to the obtained experimentally practical reinforcement absence for the indicated nanocomposite.

CONCLUSION

Thus, the obtained in the present work results have shown that the nanotubes geometry, characterized by their length, diameter and ring-like structures radius, is nanocomposites polymer/CNT-specific feature. This factor plays a crucial role in the interfacial adhesion polymer matrix − nanofiller level determination and, as consequence, in polymer nanocomposites, filled with CNT, mechanical properties formation.

REFERENCES

[1] Yanovskii Yu. G. *Nanomechanics and Strength of Composite Materials.* Moscow, Publishers of IPRIM RAN, 2008, 179 p.

[2] Bridge B. J. Mater. *Sci. Lett.*, 1989, v. 8, № 2, p. 102-103.

[3] Sheng N., Boyce M.C., Parks D.M., Rutledge G.C., Abes J.I., Cohen R.E. *Polymer*, 2004, v. 45, № 2, p. 487-506.

[4] Mikitaev A.K., Kozlov G.V., Zaikov G.E. *Polymer Nanocomposites: Variety of Structural Forms and Applications.* New York, Nova Science Publishers, Inc., 2008, 319 p.

[5] Zhirikova Z.M., Kozlov G.V., Aloev V.Z. Mater. of VII Intern. *Sci.-Pract. Conf. "New Polymer Composite Materials."* Nal'chik, KBSU, 2011, p. 158-164.

[6] Kozlov G.V., Burya A.I., Lipatov Yu.S. *Mekhanika Kompozitnykh Materialov*, 2006, v. 42, № 6, p. 797-802.

[7] Van Damme H., Levitz P., Bergaya F., Alcover J.F., Gatineau L., Fripiat J.J. *J. Chem. Phys.,* 1986, v. 85, № 1, p. 616-625.

[8] Yanovskii Yu.G., Kozlov G.V., Karnet Yu.N. *Mekhanika Kompozitsionnykh Materialov i Konstruktsii,* 2011, v. 17, № 2, p. 203-208.

In: Advanced Nanotube and Nanofiber Materials   ISBN: 978-1-62081-170-2
Editors: A. K. Haghi and G. E. Zaikov   © 2012 Nova Science Publishers, Inc.

Chapter 7

USE OF ELECTROSPINNING TECHNIQUE IN PRODUCTION OF CHITOSAN/CARBON NANOTUBE

***A. K. Haghi**[*]*

University of Guilan, Iran

1. INTRODUCTION

Over the recent decades, fabrication of polymer nanofibers have been used in many biomedical applications such as tissue engineering, drug delivery, wound dressing, enzyme immobilization, etc. [1]. The nanofiber fabrications have unique characteristics such as very large surface area, ease of functionalisation for various purposes and superior mechanical properties. The electrospinning with simple process is an important technique, which can be used for the production of polymer nanofibers with diameter in the range from several micrometers down to ten of nanometers. In electrospinning, the charged jets of a polymer solution, which are collected on a target, are created by using an electrostatic force. Many parameters can influence in quality of fibers including the solution properties (polymer concentration, solvent volatility and solution conductivity), governing variables (flow rate, voltage, and distance between tip-to-collector), and ambient parameters (humidity, solution temperature, and air velocity in the electrospinning chamber) [2].

---

[*] E-mail: Haghi@Guilan.ac.ir

In recent years, scientists have become interested in the electrospun of natural materials such as collagen [3,4], fibrogen [5], gelatin [6], silk [7], chitin [8] and chitosan [9,10] because of high biocompatible and biodegradable properties. Chitin is the second-most abundant natural polymer in the world and Chitosan (poly-(1-4)-2-amino-2-deoxy-β-D-glucose) is the deacetylated pro-duct of chitin [11]. Researchers are interested in this natural polymer because of properties, including its solid-state structure and the chain conformations in the dissolved state [12]. The chitosan/carbon nanotube composite can be produced by the hydrogen bonds due to hydrophilic positively charged polycation of chitosan due to amino groups and hydrophobic negatively charged of carbon nanotube due to carboxyl and hydroxyl groups.

This chapter discusses electrospinning of chitosan/carbon nanotube dispersion. The SEM images show homogenous chitosan/carbon nanotube nanofiber with a mean diameter of 455 nm.

2. EXPERIMENTAL

2.1. Materials

- Chitosan polymer (degree of deacetylation of 85% and molecular weight of $5{\times}10^5$) supplied by Sigma-Aldrich;
- The multi-walled carbon nanotube used in this study, supplied by Nutrino, has an average diameter of 4 nm, with purity of about 98%.

2.2. Electrospinning of Chitosan/carbon Nanotube Dispersion

Multi-walled carbon nanotube was sonicated for 10 min in solvent and then stirred for 24 hr. About 3 ml of chitosan/carbon nanotube dispersion was placed into a 5 ml syringe with a stainless steel needle having an inert diameter of 0.6 mm and was connected to positive electrode. An aluminum foil, used as the collector screen, was connected to the ground. A high-voltage power supply Gamma High Voltage Researcher ES30P-5W generated DC voltages in the range of 1-25 kV. The voltage and tip-to-collector distance were fixed at 18-24 kV and 4-10 cm, respectively. The electrospinning experiments were performed at room temperature.

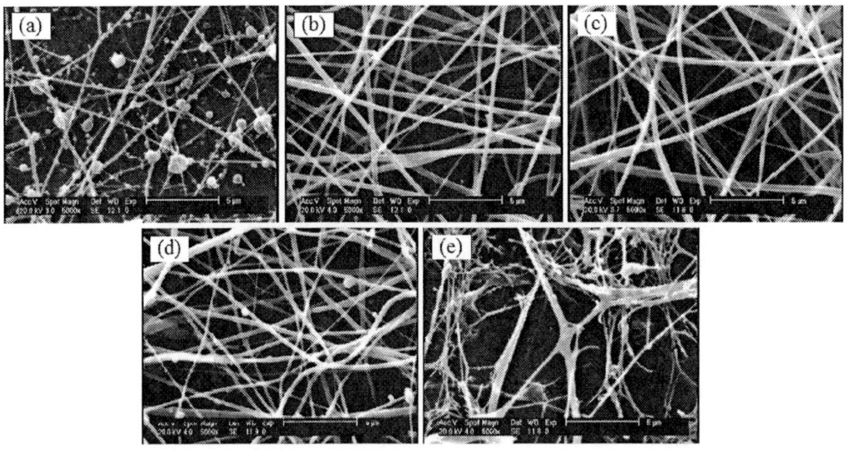

Figure 1. Scanning electron micrographs of electrospun fibers at different chitosan concentration (wt%): (a) 8, (b) 9, (c) 10, (d) 11, (e) 12, 24 kV, 5 cm, TFA/DCM: 70/30.

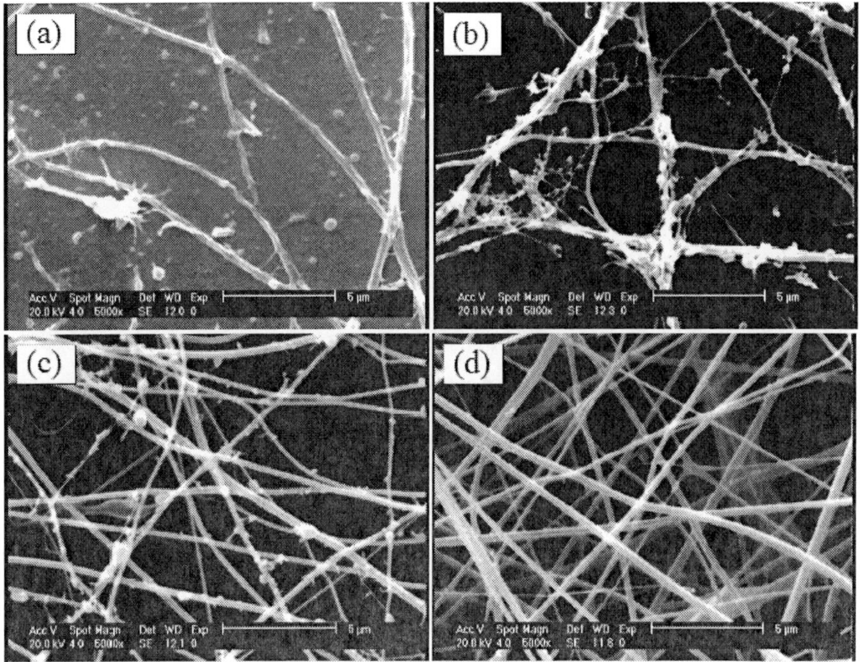

Figure 2. Scanning electronic micrographs of electrospun fibers at different voltage (kV): (a) 18, (b) 20, (c) 22, (d) 24, 5 cm, 10 wt%, TFA/DCM: 70/30.

3. RESULTS AND DISCUSSION

The different solvents including acetic acid 1-90%, formic acid, and TFA/DCM tested for the electrospinning of chitosan/carbon nanotube. No jet was seen upon applying the high voltage even above 25 kV by using of acetic acid 1-30% and formic acid as the solvent for chitosan/carbon nanotube. When the acetic acid is 30-90%, used as the solvent, beads were deposited on the collector. Therefore, under these conditions, nanofiber of carbon nanotube/chitosan was not obtained.

Figure 1 shows Scanning electronic micrographs of the carbon chitosan/nanotube electrospun fibers in different concentration of chitosan in TFA/DCM (70:30) solvent. As presented in Figure 1a, at low concentrations of chitosan, the beads deposited on the collector, and thin fibers coexited among the beads. As the concentration of chitosan increased (Figures 1a-c), the beads decreased significantly. Figure 1c show homogenous electrospun fibers with minimum beads, thin fibers and interconnected fibers. Increase of chitosan concentration leads to increase of interconnected fibers as shown in Figures 1d-e. The average diameter of chitosan/carbon nanotube fibers were increased by increasing concentration of chitosan (Figures 1 a-e). Hence, chitosan/carbon nanotube solution of TFA/DCM (70:30) with 10 wt% of chitosan resulted in optimized conditions for electrospinning of this solution with an average diameter of 455 nm (Figure 1c: with diameter distribution of, 306-672).

When the voltage was low, the beads were deposited on collector (Figure 2a). As shown in Figures 2a-d, the number of beads decreased by increasing the voltage from 18 kV to 24 kV. In our study, the average diameter of fibers prepared by 18 kV measured as 307 nm. As the applied voltage increased, the average fiber diameters increased as well. The average diameter of fibers for 20 kV (2b), 22 kV (2c), and 24 kV (2d), was 308 (194-792), 448 (267-656), 455 (306-672) respectively.

The morphologies of chitosan/carbon nanotube electrospun fibers at different distance tip-to-collector are presented in Figure 3. When the distance tip-to-collector was low, the solvent did not vapor, hence a little interconnected fiber (with high fiber diameter) deposited on the collector (as shown in Figure 3a). In 5 cm distance tip-to-collector (Figure 3b), more homogenous fibers with negligible beads were obtained. However, the beads increased by increasing of distance tip-to-collector (Figure 3b to Figure 3f). Also, our study represented that the diameter of electrospun fibers decreased by increasing of distance tip-to-collector (as shown in Figures 3b, 3c, 3d, we have; 455 (306-672), 134 (87-163), 107 (71-196)). The fibers prepared within a distance of 8 cm (Figure 3e) and 10 cm (Figure 3f), the defects and non-homogenous diameter fibers were remarkable. However, a distance of 5 cm for tip-to-collector seems to be reliable for electrospinning.

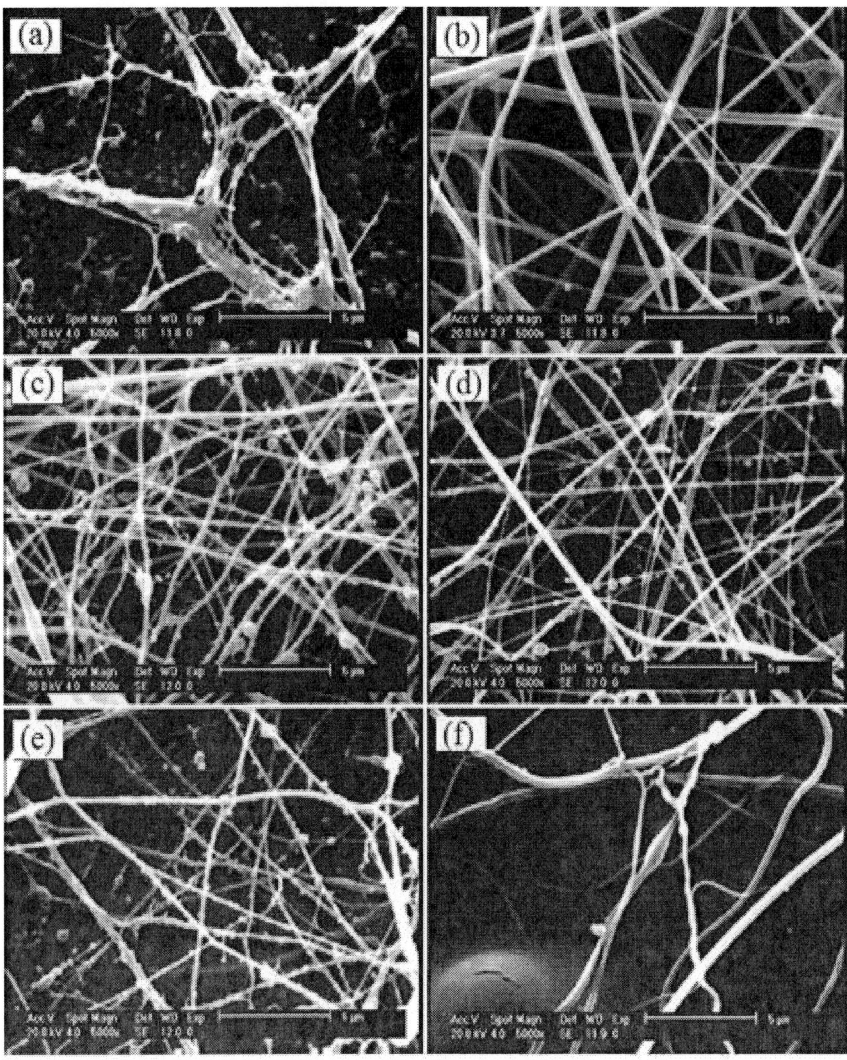

Figure 3. Scanning electronic micrographs of electrospun fibers of Chitosan/Carbon nanotubes at different tip-to-collector distances (cm): (a) 4, (b) 5, (c) 6, (d) 7, (e) 8, (f) 10, 24 kV, 10 wt%, TFA/DCM: 70/30

CONCLUSION

Several solvents including acetic acid 1-90%, formic acid, and TFA/DCM (70:30) were used for electrospinning of chitosan/carbon nanotube dispersion. It is observed that the TFA/DCM (70:30) solvent was the only solvent with a proper reliability for electrospinnability of chitosan/carbon nanotube. This is a significant improvement in electrospinning of chitosan/carbon nanotube dispersion. It is also observed that the homogenous fibers with an average diameter of 455 nm (306-672) could be prepared with chitosan/carbon nanotube dispersion in TFA/DCM 70:30. Meanwhile, the SEM images showed that the fiber diameter decreased by decreasing of voltage and increasing the distance of tip-to-collector.

REFERENCES

[1]   M. Ziabari, V. Mottaghitalab, A. K. Haghi, Simulated image of electrospun nonwoven web of PVA and corresponding nanofiber diameter distribution, *Korean Journal of Chemical engng*, Vol.25, No. 4, pp. 919-922,2008.

[2]   M. Ziabari, V. Mottaghitalab, A. K. Haghi, Evaluation of electrospun nanofiber pore structure parameters, *Korean Journal of Chemical engng*, Vol.25, No. 4, pp. 923-932,2008.

[3]   M. Ziabari, V. Mottaghitalab, A. K. Haghi, Distance transform algorithm for measuring nanofiber diameter, *Korean Journal of Chemical engng*, Vol25, No. 4, pp. 905-918,2008.

[4]   Kyong Su Rho, Lim Jeong, Gene Lee, Byoung-Moo Seo, Yoon Jeong Park, Seong-Doo Hong, Sangho Roh, Jae Jin Cho, Won Ho Park, Byung-Moo Min, Electrospinning of collagen nanofibers: Effects on the behavior of normal human keratinocytes and early-stage wound healing, *Biomaterials* 27 (2006) 1452–1461.

[5]   Michael C. McManus, Eugene D. Boland, David G. Simpson, Catherine P. Barnes, Gary L. Bowlin, Electrospun fibrinogen: Feasibility as a tissue engineering scaffold in a rat cell culture model, *InterScience*. DOI: 10.1002/jbm.a.30989.

[6]   Zheng-Ming Huang, Y.Z. Zhang, S. Ramakrishna, C.T. Lim, Electrospinning and mechanical characterization of gelatin nanofibers, *Polymer* 45 (2004) 5361–5368.

[7]   Xiaohui Zhang, Michaela R. Reagan, David L. Kaplan, Electrospun silk biomaterial scaffolds for regenerative medicine, *Advanced Drug Delivery Reviews* 61 (2009) 988–1006.

[8]   Hyung Kil Noh, Sung Won Lee, Jin-Man Kim, Ju-Eun Oh, Kyung-Hwa Kim, Chong-Pyoung Chung, Soon-Chul Choi, Won Ho Park, Byung-Moo Min, Electrospinning of chitin nanofibers: Degradation behavior and cellular response to normal human keratinocytes and fibroblasts, *Biomaterials* 27 (2006) 3934–3944.

[9]   Kousaku Ohkawa, Ken-Ichi Minato, Go Kumagai, Shinya Hayashi, and Hiroyuki Yamamoto, Chitosan Nanofiber, *Biomacromolecules* 2006, 7, 3291-3294.

[10] Xinying Geng, Oh-Hyeong Kwon, Jinho Jang, Electrospinning of chitosan dissolved in concentrated acetic acid solution, *Biomaterials* 26 (2005) 5427–5432.

[11] O. C. Agboh and Y. Qin, Chitin and Chitosan Fibers, Polymers for Advanced Technologies 8 (1997) 355–365. [12] R. Marguerite, Chitin and chitosan: properties and applications, *Prog. Polym. Sci. 31* (2006) 603-632.

In: Advanced Nanotube and Nanofiber Materials   ISBN: 978-1-62081-170-2
Editors: A. K. Haghi and G. E. Zaikov   © 2012 Nova Science Publishers, Inc.

Chapter 8

COMBUSTION AND THERMAL DEGRADATION OF POLYPROPYLENE IN THE PRESENCE OF MULTI-WALLED CARBON NANOTUBE COMPOSITES

G. E. Zaikov[1], S. M. Lomakin[1], N. G. Shilkina[2] and R. Kozlowski[3]

[1] NM Emanuel Institute of Biochemical Physics of Russian
Academy of Sciences, Moscow, Russia
[2] NN Semenov Institute of Chemical Physics of Russian
Academy of Sciences, Moscow, Russia
[3] Institut Inzynierii Materialow Polimerowych I Barwnikow,
Torun, Poland

INTRODUCTION

At present time, great attention is given to the study of properties of polymeric nanocomposites produced on the basis of well-known thermoplastics (PP, PE, PS, PMMA, polycarbonates, polyamides) and carbon nanotubes (CN). CNs are considered to have the wide set of important properties like thermal stability, reduced combustibility, electroconductivity, etc. [1-7]. Thermoplastic polymer nanocomposites are generally produced with the use of melting technique [1-12].

Development of synthetic methods and the thermal characteristics study of PP/multi-walled carbon nanotube (MWCNT) nanocomposites were taken as an objective in this paper.

A number of papers pointed at synthesis and research of thermal properties of nanocomposites (atactic polypropylene (aPP)/MWCNT) were reported [10-12]. It is remarkable that PP/MWCNT composites with minor level of nanocarbon content (1-5% by weight) were determined to obtain an increase in thermal and thermal-oxidative stability in the majority of these publications.

Thermal stability of aPP and aPP/MWCNT nanocomposites with the various concentrations of MWCNT was studied in the paper [10]. It was shown that thermal degradation processes are similar for aPP and aPP/MWCNT nanocomposites, and initial degradation temperatures are the same. However, the maximum mass loss rate temperature of PP/MWCNT nanocomposites with 1 and 5% wt of MWCNT raised by 40° - 70°C as compared with pristine aPP.

Kashiwagi et al. published the results of study of thermal and combustion properties of PP/MWCNT nanocomposites [11, 12]. A significant decrease of maximum heat release rate was detected during combustion research with use of cone calorimeter. A formation of char network structure during the combustion process was considered to be the main reason of combustibility decrease. The carbonization influence upon combustibility of polymeric nanocomposites was widely presented in literature [10-12, 13]. Notably, Kashiwagi et al. [11, 12] were the first to hypothesize that abnormal dependence of maximum heat release rate upon MWCNT concentration is closely related with thermal conductivity growth of PP/MWCNT nanocomposites during high-temperature pyrolysis and combustion.

EXPERIMENTAL

Materials

Isotactic polypropylene (melting flow index = 0.7 g/10 min) was used as a polymer matrix in this paper. Multi-walled carbon nanotubes (MWCNT) (purchased from Shenzhen Nanotechnologies Co. Ltd.) were used as a carbon-containing nanofillers. This product contains low amount of amorphous carbon (less than 0.3 wt%) and could be produced with different size characteristics—

different length and different diameter and therefore different diameter to length ratio. Size characteristics for three MWCNT used in this paper are given in Table 1. Sizes and structure of initial MWCNT were additionally estimated by SEM (Fig.1).

Table 1. Properties of MWCNT

Designation	D, nm	L, μm	Density, g/cm^3	Specific surface area, m^2/g
MWCNT (K1)	<10	5-15	2	40 - 300
MWCNT (K2)	40-60	1-2	2	40 - 300
MWCNT (K3)	40-60	5-15	2	40 - 300

Nanocomposite Processing

Compositions were prepared by blending carbon nanotubes with melted polymer in a laboratory mixer Brabender at 190°C. TOPANOL® (1,1,3-tris (2-methyl-4-hydroxy-5-t-butylphenyl) butane), and DLTP (dilaurylthiodipropionate) were added in the amount of 0.3 and 0.5 wt% as antioxidants to prevent thermal-oxidative degradation during polymer processing.

A number of different covalent and non-covalent nanotube modifications (organofillization) were reported to be used to achieve greater structure similarity and therefore greater nanotube distribution in a polymer matrix [14-23]. In order to functionalize MWCNT, we used preliminary ozone treatment of MWCNT followed by ammonolysis of epoxy groups on the MWCNT surface. The selective ozonization of MWCNT was carried out with ozone-oxygen mixture (ozone concentration was 2.3×10^{-4} mol/L) in a bubble reactor. Then the ammonolysis of oxidized MWCNT has been carried out by *tert*-butylamine in the ultrasonic bath (35 kHz) at 50°C for 120 min with following evaporation of *tert*-butylamine excess. IR transmission spectra of tablet specimens of MWCNTs in KBr matrix was analyzed by using Perkin-Elmer 1725X FTIR spectrometer, and the presence of the alkylamine groups at the MWCNT surface was confirmed by the appearance of the characteristic band ~1210 cm^{-1} corresponding to the valency vibration of the bond –C–N←.

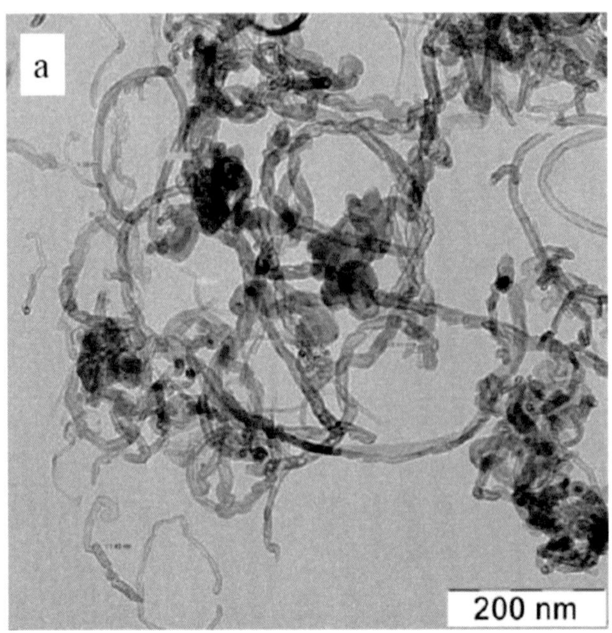

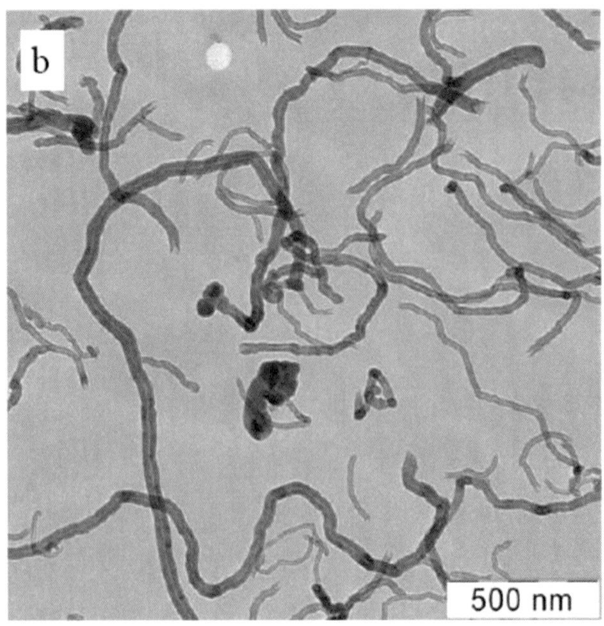

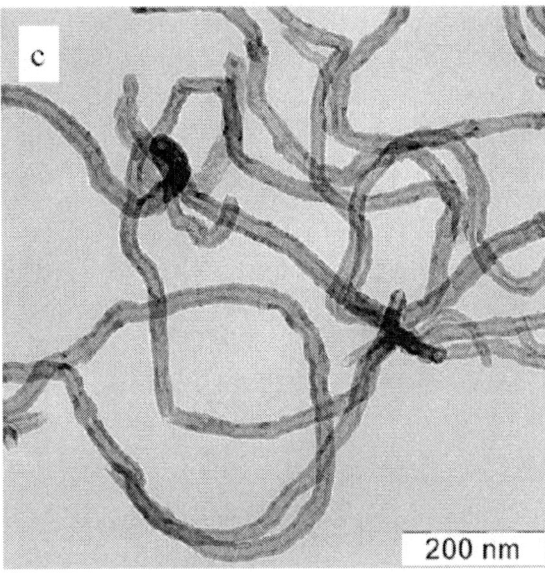

Figure 1. SEM images of original MWCNTs: (a) MWCNT(1); (b) MWCNT(2); (c) MWCNT(3).

Investigation Techniques

Scanning electron microscopy (SEM). The degree of MWCNT distribution in polymer matrix was analyzed with scanning electron microscope JSM-35. Low-temperature chips derived from film-type samples were used for this analysis.

Transmission electron microscopy (TEM). The degree of nanotube dispersion in polymer matrix was studied with transmission electron microscopy (LEO912 AB OMEGA, Germany). Microscopic sections with 70 − 100 nm width prepared with ultramicrotome "Reichert − Jung Ultracut" with diamond cutter at -80°C. Microscopic analysis was made with accelerating potential of about 100 kV without chemical sample staining.

Thermogravimetric analysis (TG). A NETZSCH TG 209 F1 Iris thermomicrobalance has been employed for TGA measurements in oxidizing (oxygen) atmosphere. The measurements were carried out at a heating rate of 20 K/min.

Combustibility characteristics (cone calorimeter) were performed according to the standard procedures ASME E1354/ISO 5660 using a DUAL

CONE 2000 cone-calorimeter (FTT). An external radiant heat flux of 35 kW/m^2 was applied. All of the samples having a standard surface area of 70×70 mm and identical masses of 13.0±0.2g were measured in the horizontal position and wrapped with thin aluminum foil except for the irradiated sample surface.

Heat capacity and heat conductivity were determined with the use of NETZSCH 457 MicroFlash.

The electron paramagnetic resonance spectroscopy (EPR) measurements were performed in air with the PP/MWCNT (10 wt%) samples using a Mini-EPR SPIn Co. Ltd spectrometer with 100 kHz field modulation. The g factor and EPR intensity (X-band) were measured with respect to a standard calibrating sample of Mn^{2+} and ultramarine.

RESULTS AND DISCUSSION

Nanocomposite Structure

Dispersion analysis of MWNT in nanocomposites. PP/MWNT nanocomposites with original and modified MWNT were produced. Filler concentration varied from 1 to 7 wt% weight percent (0.5-3.5 volume percent correspondingly). Distribution pattern for composites with modified and non-modified nanotubes was studied with TEM methods (Fig 2). According to TEM images, the addition of 1% by weight leads to sufficiently uniform distribution. However, agglomeration of nanotubes was detected for more concentrated nanocomposites, especially for PP/MWNT with MWNT average diameter less than 10 nm (K1).

TEM images for PP nanocomposites with 5 wt% modified and non-modified MWNT (2.5% by volume) are shown in Fig.2. According to Fig. 2, it could be stated that modified nanotubes (K2 and K3) used during melding process are present as individual particles in nanocomposite in most cases. The number and size of agglomerates is reduced due to increased organophility and improved thermodynamic compatibility with nonpolar polymer.

However, preliminary modification does not lead to uniform filler distribution for nanocomposites containing thin nanotubes (K1). This could be explained by the fact that interaction energy of CNT is more dependent on nanotube diameter than on its length. Molecular dynamic computation given in paper [24] showed that blending polymers with nanotubes becomes more thermodynamically favorable with increase of nanotubes diameter, owing to the fact that cohesion energy is decreased between the nanotubes and remains almost the same between nanotubes and polymer.

Thus, mixing the thinnest nanotubes (K1) with PP leads to inevitable nanotube agglomeration in nanocomposite sample volume. Nanotube surface modification **used in this paper didn't result in complete overcome** of nanotube agglomeration tendency for K1 nanotubes [25]. Therefore, the PP/MWNT(K3) nanocomposites presented the main subjects of inquiry in the present study.

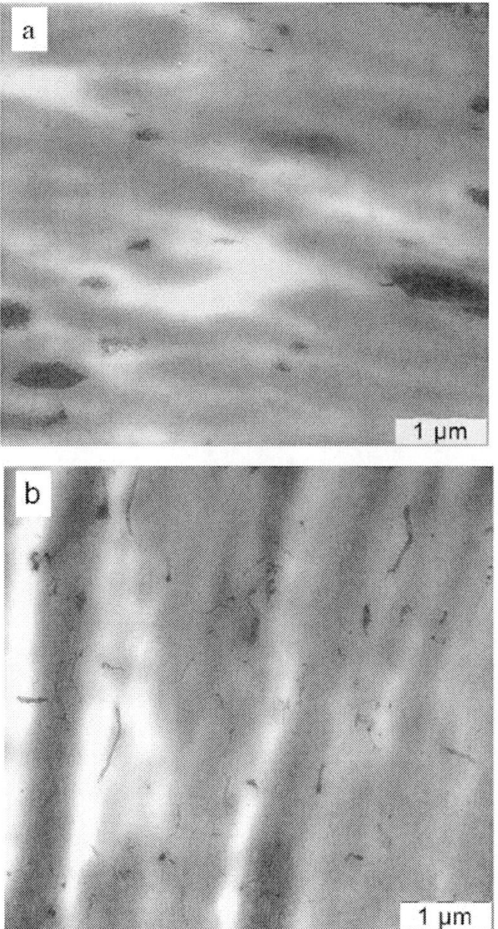

Figure 2. (Continued).

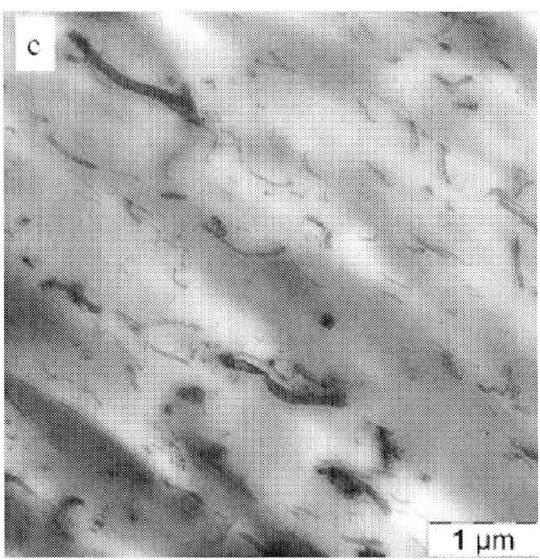

Figure 2. TEM images of PP/MWCNT nanocomposites showing dispersion of MWCNT in a polymer matrix: (a) PP/MWCNT(1); (b) PP/MWCNT(2); (c) PP/MWCNT(3).

Thermal-Oxidative Degradation of PP/MWCNT Nanocomposites

The diverse behavior of PP and PP/MWCNT nanocomposites with 1, 3 and 5 wt% of MWCNT(3). (Fig. 3) shows that the influence of MWCNTs on the thermal-oxidation process resulted in higher thermal-oxidative stability of PP/MWCNT nanocomposites. It is possible to see a regular increase in the temperature values of the maximum mass loss rates (up to 60°C) for the PP/MWCNT as compared to pristine PP (Fig. 3).

Detailed analysis of TGA graphs (Fig. 3) allows claiming that thermal stability increase is achieved even by addition of 1 wt% of MWCNT to PP, while further addition doesn't lead to such fundamental growth. In addition, Fig. 4 shows the comparative results for onset degradation temperatures ($T_{on.}$) and the maximum mass loss temperatures (T_{max}) of PP/MWCNT nanocomposites with the different types and concentrations of MWCNT. One can see nonlinear relation of ($T_{on.}$) and (T_{max}) vs. MWCNT concentration in the PP compositions (Fig. 4).

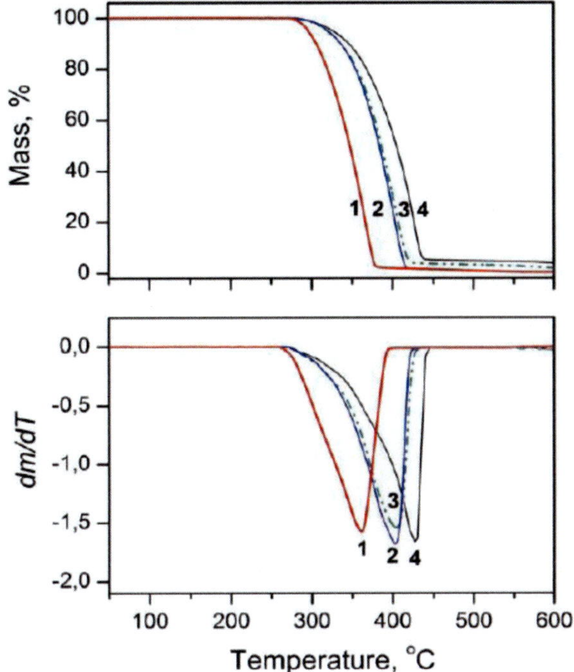

Figure 3. TG and DTG curves for PP (1) and PP/MWCNT(3) composites with 1 (2), 3 (3) and 5 wt% (4) filler loadings.

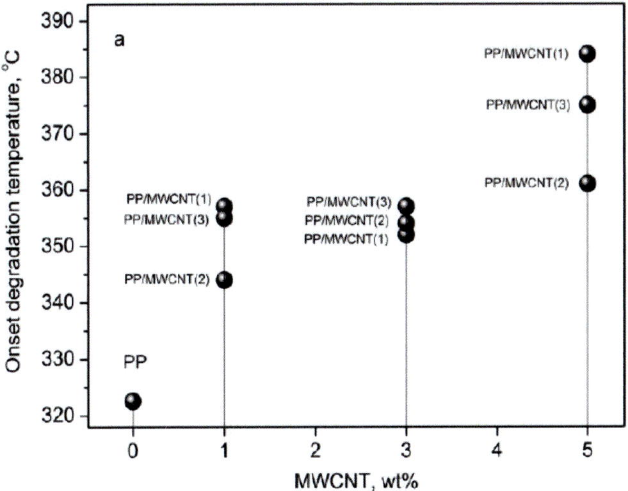

Figure 4. (Continued).

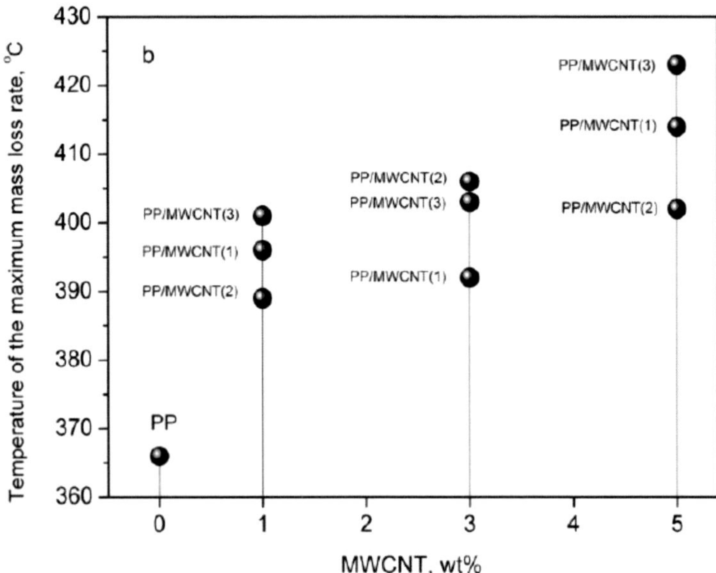

Figure 4. Comparative diagrams showing the onset degradation temperatures (a) and the maximum mass loss temperatures (b) for PP and PP/MWCNT nanocomposites with the different types and concentrations of MWCNT.

At the present time, the nature of thermal stability effect caused by MWCNT addition to polymers is an object of comprehensive study. Most likely, MWCNTs could be considered as high temperature stabilizers (antioxidants) in reactions of thermal-oxidative degradation by analogy with fullerenes [26]. Stabilizing effect caused by addition of MWCNT was previously detected for PP/MWCNT nanocomposites [8]: the temperature of the maximum mass loss rates of PP/MWCNT (9 wt%) was increased by 50°C as compared with pristine PP.

Results achieved in this study confirm the previous findings of the inhibiting effect of MWNCT upon the PP/MWCNT nanocomposite thermal-oxidative degradation. Obviously, the complex nature of this effect is closely related to radical-acceptor properties of MWCNT resulting in chain termination reactions of alkyl/alkoxyl radical, which lead to cross-linking and carbonization process in PP/MWCNT nanocomposites. Carbonization phenomenon was reported previously in papers aimed at PP/MWCNT heat-resistance and flame-retardancy study [11, 12].

Kinetic Analysis of Thermal Degradation of PP/MWNT

Kinetic studies of material degradation have a long history, and there exists a long list of data analysis techniques employed for the purpose. Often, TGA is the method of choice for acquiring experimental data for subsequent kinetic calculations, and this technique was employed here. It is commonly accepted that the degradation of materials follows the base equation (1) [27]

$$dc/dt = - F(t, T, c, p) \tag{1}$$

where: t - time, T - temperature, c_o - initial concentration of the reactant, and p - concentration of the final product. The right-hand part of the equation $F(t, T, c, p)$ can be represented by the two separable functions, $k(T)$ and $f(c, p)$:

$$F(t, T, c, p) = k [T(t){\cdot}f(c, p)] \tag{2}$$

Arrhenius equation (4) will be assumed to be valid for the following:

$$k(T) = A{\cdot}exp(-E/RT) \tag{3}$$

Therefore,

$$dc/dt= - A{\cdot}exp(-E/RT){\cdot}f(c,p) \tag{4}$$

All feasible reactions can be subdivided onto classic homogeneous reactions and typical solid-state reactions, which are listed in Table 2. The analytical output must provide good fit to measurements with different temperature profiles by means of a common kinetic model.

Thermogravimetric analysis of PP and PP nanocomposite degradation was carried out in dynamic conditions at the rates of 2.5, 5 and 10 K/min on air.

Model-independent estimation of activation energy using Friedman approach [28] was taken to get preliminary model analysis for thermal degradation and selection of initial conditions. According to this evaluation, a two-step process ($A{\rightarrow}X_1{\rightarrow}B{\rightarrow}X_2{\rightarrow}C$) was chosen for PP degradation. Taking into account the carbonization stage, the more complex three-step process ($A{\rightarrow}X_1{\rightarrow}B{\rightarrow}X_2{\rightarrow}C{\rightarrow}X_3{\rightarrow}D$) was selected for PP/MWNT degradation [27, 29].

According to the results of nonlinear regression and taking the set of reaction models into consideration, we computed the values of active kinetic parameters, which represent the best approximation of experimental TGA graphs (Fig. 5, Table 3).

Table 2. Considered reaction models $dc/dt=-A \cdot exp(-E/RT)f(c, p)$

Reaction models	$f(c, p)$
First order (F_1)	c
Second order (F_2)	c^2
n-order (F_n)	c^n
Two-dimensional phase boundary (R_2)	$2 \cdot c^{1/2}$
Three-dimensional phase boundary (R_3)	$3 \cdot c^{2/3}$
One-dimensional diffusion (D_1)	$0.5/(1 - c)$
Two-dimensional diffusion (D_2)	$-1/ln(c)$
Three-dimensional diffusion, Jander's type (D_3)	$1.5c^{1/3}(c^{-1/3} - 1)$
Three-dimensional diffusion, Ginstling-Brounstein (D_4)	$1.5/(c^{-1/3} - 1)$
One-dimensional diffusion (Fick law) (D_{1F})	-
Three-dimensional diffusion (Fick law) (D_{3F})	-
Prout-Tompkins equation (B_1)	$c \cdot p$
Expanded Prout-Tompkins equation (B_{na})	$c^n \cdot p^a$
First-order reaction with autocatalysis by X (C_{1-X})	$c \cdot (1+K_{cat} X)$
n-order reaction with autocatalysis by X (C_{n-X})	$c^n \cdot (1+K_{cat} X)$
Two-dimensional nucleation, Avrami-Erofeev equation (A_2)	$2 \cdot c \cdot (-ln(c))^{1/2}$
Three-dimensional nucleation, Avrami-Erofeev equation (A_3)	$3 \cdot c \cdot (-ln(c))^{2/3}$
n- dimensional nucleation, Avrami-Erofeev equation (A_n)	$n \cdot c \cdot (-ln(c))^{(n-1)/n}$

Table 3. Kinetic parameters for thermal degradation of (a) PP (Fn → Fn) and (b) PP/MWNT nanocomposite (Fn →D1→ Fn). TGA analysis was performed in air flow with the use of multiple non-linear regression analysis for model processes

(a)

Reaction model	Kinetic parameters	Values	Correlation coefficient
Fn →Fn	$logA_1$, s^{-1}	9.53	0.9996
	E_1, kJ/mol	110.25	
	n_1	1.89	
	$logA_2$, s^{-1}	15.25	
	E_2, kJ/mol	150.65	
	n_2	1.50	

(b)

Reaction model	Kinetic parameters	Values	Correlation coefficient
$F_n \rightarrow D1 \rightarrow F_n$	$\log A_1$, s^{-1}	6.3	0.9996
	E_1, kJ/mol	105.1	
	n_1	0.91	
	$\log A_2$, s^{-1}	7.4	
	E_2, kJ/mol	120.4	
	$\log A_3$, s^{-1}	16.7	
	E_3, kJ/mol	229.5	
	n_3	0.5	

Two-step PP thermal-oxidative degradation in dynamic heating conditions was confirmed by obtained data [30]. At the first stage, the values of activation energy and pre-exponential factor are 110.25 kJ/mol and $10^{9.5}$ s^{-1} correspondingly, while the reaction order is close to 2 (1.89). The values of activation energy and pre-exponential factor are larger on the second stage (E_2 = 150.65 kJ/mol, $A_2 = 10^{15.3}$ s^{-1}) with effective reaction order of $n_2 = 1.50$.

The preferred model for PP/MWNT thermal-oxidative degradation and with respect to statistical analysis of kinetic parameters is composed of three consecutive reactions $F_n \rightarrow D_1 \rightarrow F_n$, where D_1 – one-dimensional diffusion, and F_n – n-order reaction (Fig 4b, Table 3b). In this case, the first step activation energy is equal to 105.1 kJ/mol, and reaction order is close to 1 ($n_1 = 0.91$). On the second step, which is described as one-dimensional diffusion, the value of activation energy is equal 120.4 kJ/mol, while the value is almost twice large for the third step ($E_3 = 229.5$ kJ/mol) with effective reaction order of $n_3 = 0.5$ (Table 3b).

Comparison of thermal oxidative degradation parameters for PP/MWNT with layered silicate PP//MMT showed that the values of activation energy of the second and the third stages are higher for PP/MWNT:

E_2=120.4 kJ/mol and E_3=229.5 kJ/mol for PP/MWNT;

E_2=100.0 kJ/mol and E_3=199.8 kJ/mol for PP/MPP/MMT correspondingly [30].

This data may testify to more intensive carbonization in case of PP/MWNT than in case of PP/MMT, which finally leads to decrease in RHR value.

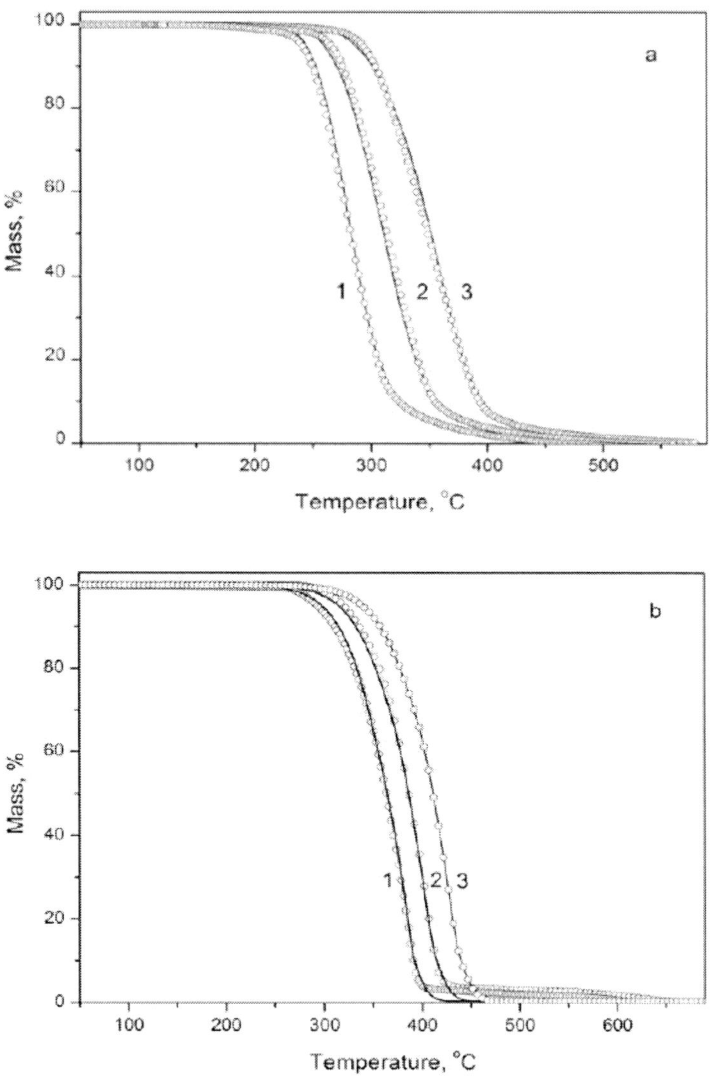

Figure 5. Nonlinear kinetic modeling of a – PP and b – PP/MWCNT(3) thermal-oxidative degradation in air. Comparison between experimental TG data (dots) and the model results (firm lines) at several heating rates: (1) – 2.5, (2) – 5, (3) – 10 K/min.

COMBUSTIBILITY OF PP/MWCNT NANOCOMPOSITES

Figure 6 depicts the plots of the heat release rate (RHR), as basic flammability characteristic, vs. time for PP, as well as for the PP/MWCNT nanocomposites.

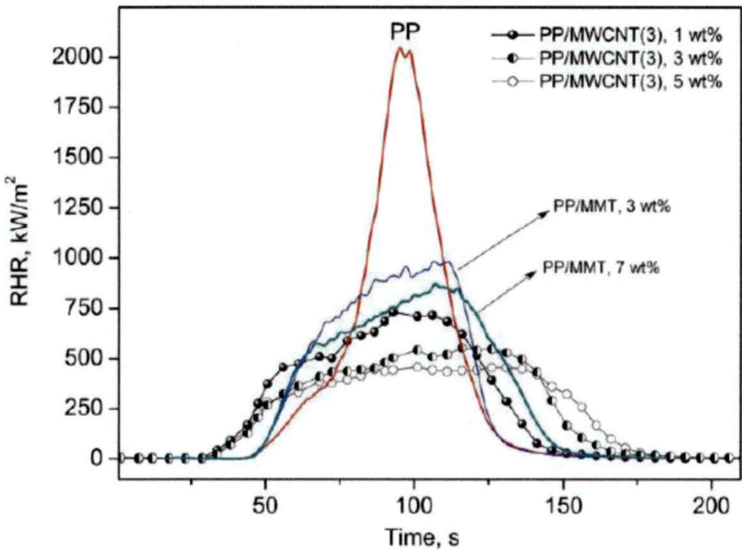

Figure 6. Rate of heat release vs. time for PP, PP/MWCNT and PP/MMT(Cloisite 20A) nanocomposites obtained by cone calorimeter at the incident heat flux of 35 kW m^{-2}.

From Fig. 6, it could be seen that the maximum heat release rate for pristine PP is 2076 kW/m^2, whereas that for the PP/MWCNT % nanocomposites (1 wt%), PP/MCWNT (3 wt%) and the PP/MCWNT (5 wt.%) RHR values are 729 kW/m^2, 552.8 kW/m^2, and 455.8 kW/m^2, respectively; thus, the peak heat release rate decreases by 65%, 73% and 78%.

The observed flame-retardancy effect is associated with solid-phase carbonization reactions by analogy with layered silicates [31, 32]. In early paper [30], we have found that additions of 3 and 7 wt % of layered silicate (Cloisite 20A) to the PP compositions PP lead to RHR decrease by 51 and 57% as compared with pristine PP (Fig.5).

We believe that a higher carbonization effectiveness of MWCNTs depends on their heat conductivity. It is well known that PP has a low thermal conductivity at standard conditions and is characterized by a minor increase with temperature up to melting point (~0.2 W/m K). On the other hand, the heat conductivity of individual MWCNT is extremely high and equals 3000 W/m K [33, 34]. During the high temperature pyrolysis of PP/MWCNT composition at temperatures above 300-400°C, the heat conductivity can rise up to 20W/m K [34] due to actual increase of MWCNT concentration in composition caused by volatilization of polypropylene degradation products (Fig.7). The induced heating of PP/MWCNT intensifies a steady carbonization and charring of the samples and leads to decrease of RHR peak value (Fig. 6).

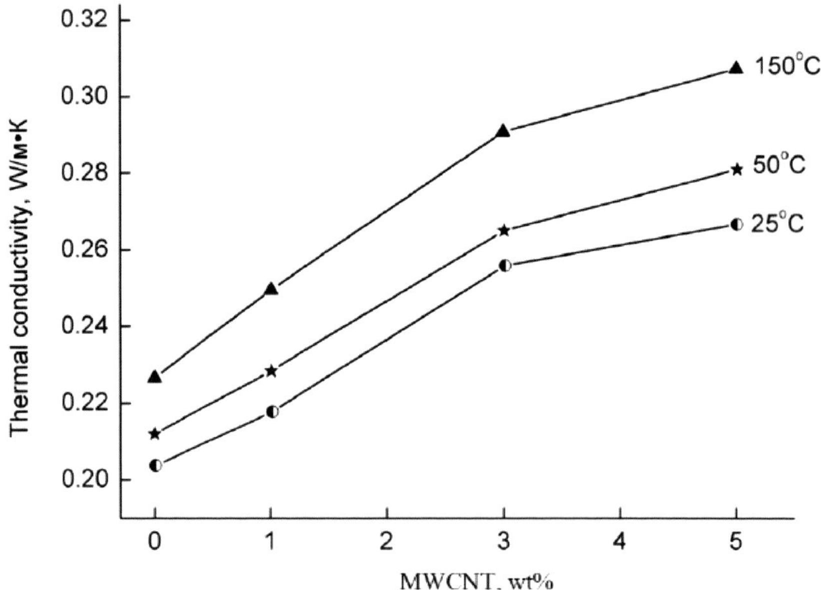

Figure 7. Temperature dependence of the thermal conductivity of PP/MWCNT (3) nanocomposite with different loadings of MWCNT.

Figures 8 and 9 show graphs for the specific extinction area and effective heat of combustion, correspondingly, for PP and PP/MWCNT(3) nanocomposites. Calculated values of effective heat of combustion for PP and PP/MWNT demonstrate invariant shift of this parameter for these nanocomposites.

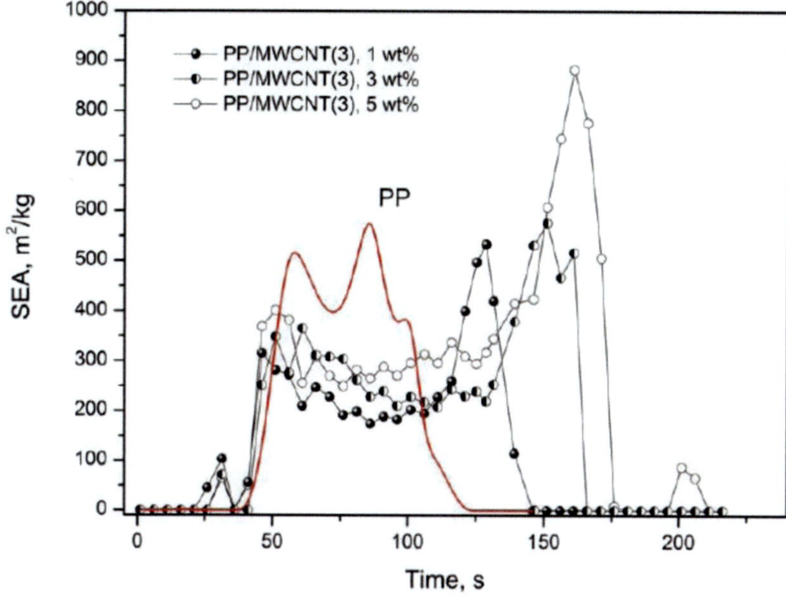

Figure 8. Specific extinction area vs. time for PP and PP/MWCNT(3) nanocomposites obtained by cone calorimeter at the incident heat flux of 35 kW m^{-2}.

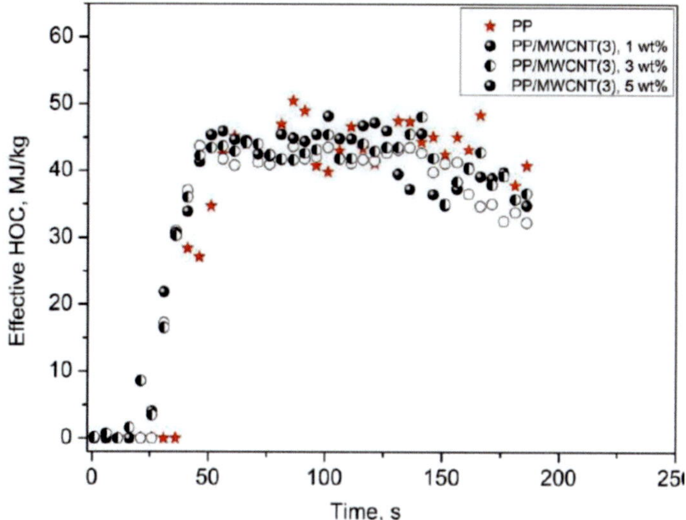

Figure 9. Effective heat of combustion vs. time for PP and PP/MWCNT(3) nanocomposites obtained by cone calorimeter at the incident heat flux of 35 kW m^{-2}.

In the present study, EPR research were performed to follow formation of stable radicals, responsible for carbonization process, upon isothermal heating of PP/MCWNT (10 wt%) in air at 350°C.

Fig. 10a shows EPR spectrum of the stable paramagnetic centers formed in the samples of PP/MCWNT (10% wt) heating in air at 350°C. When heated in air, PP/MCWNT specimen was placed into an EPR sample tube, a narrow singlet signal with a line width of $\Delta H_{1/2} = 0{,}69$ mT and a g value of 2.003 was detected due to the stable radicals generation, analogous to those previously registered during polymers carbonization process [35]. No EPR signal similar to that of PP/MCWNT samples was observed in the samples of pristine PP and MCWNT samples heated at 350°C in air. It should be noted that although iron impurity from MWNCT has been mentioned in other studies on pyrolysis of polymer nanocomposites as the radical traps [36, 11], the EPR analyses in the current study showed the presence of paramagnetic centers relating to carbonaceous stable radicals only.

As it is seen from Fig. 10b, the formation of stabilized radicals occurs with pronounced induction period, which is related to antioxidant properties of MWNCT. Such a type of kinetic dependence is coincided with an oxygen uptake kinetics observed during inhibited polyolefines thermal oxidation. Moreover, no EPR signals were observed in the samples of the PP/MCWNT samples heated at 350°C in inert Ar.

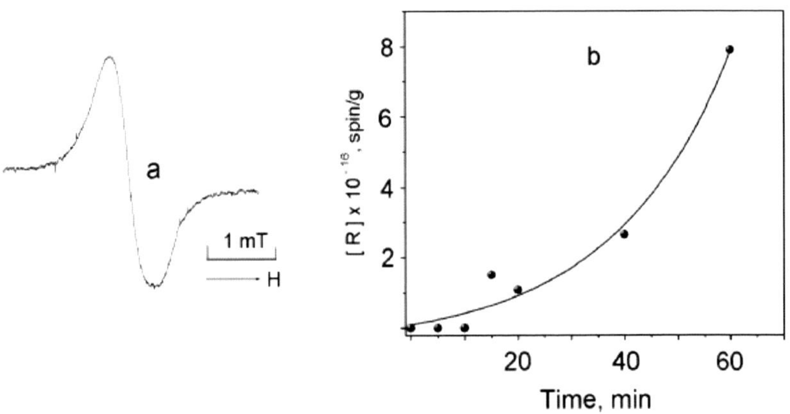

Figure 10. EPR spectrum of the stable paramagnetic centers formed in the samples of PP/MCWNT (10% wt) heating in air at 350°C – (a), kinetic dependence of stable radicals generation from PP/MCWNT (10% wt) under isothermal heating at 350°C in air – (b).

Thus, multi-walled carbon nanotubes are considered to be more effective filling agents than layered silicates in the terms of improvement of thermal properties and flame retardancy of PP matrix. This could be explained by the specific antioxidant properties and high thermal conductivity of MWCNT, which determine the carbonization reactions during thermal-oxidative degradation process.

REFERENCES

[1]   Shaffer MSP, Windle AH. *Adv. Mater.* 1999, 11, 937.

[2]   Qian D, Dickey EC, Andrews R, Rantell T. *Appl. Phys. Lett.* 2000, 76, 2868.

[3]   Jin Z, Pramoda KP, Xu G, Goh SH. *Chem. Phys. Lett.* 2001, 337, 43.

[4]   Thostenson ET, Chou TW. *J. Phys. D: Appl. Phys.* 2002, 35, L77.

[5]   Bin Y, Kitanaka M, Zhu D, Matsuo M. *Macromolecules* 2003, 36, 6213.

[6]   Potschke P, Dudkin SM, Alig I. *Polymer.* 2003, 44, 5023.

[7]   Safadi B, Andrews R, Grulke EA. *J. Appl. Polym. Sci.* 2002, 84, 2660.

[8]   P.C.P. Watts, P.K. Fearon, W.K. Hsu, N.C. Billingham, H.W. Kroto and D.R.M. Walton, *J. Mater. Chem.* 2003, 13, 491.

[9]   Watts PCP, Hsu WK, Randall DP, Kroto HW and Walton DRM., *Phys. Chem. Chem. Phys.,* 2002, 4, 5655.

[10]   Jun Yang, Yuhan Lin, Jinfeng Wang. *J. Appl. Polym. Sci.* 2005, 98,3,1087.

[11]   Kashiwagi T, Grulke E, Hilding J, Groth K,; Harris RH, Butler K., Shields JR,; Kharchenko S, Douglas J, *Polymer.* 2004, 45, 4227–4239.

[12]   Kashiwagi T, Grulke E, Hilding J, Harris Jr RH, Awad WH, Dougls J. *Macromol. Rapid Commun.* 2002, 23, 761, 5.

[13]   Lomakin SM, Novokshonova LA, Brevnov PN, Shchegolikhin AN, *Journal of Materials Science.* 2008, 43, 4, 1340.

[14]   Chen J, Hamon M A, Hu H, Chen Y, Rao A M, Eklund P C and Haddon R C *Science.* 1998, 282, 95.

[15]   Stevens J L, Huang A Y, Peng H, Chiang I W, Khabashesku V N and Margrave J L *Nano Lett.,* 2003, 3, 331.

[16]   Eitan A, Jiang K, Dukes D, Andrews R and Schadler L S *Chem. Mater.,* 2003, 15 3198.

[17]   Hu H, Ni Y, Montana V, Haddon R C and Parpura V *Nano Lett.,* 2004, 4, 507.

[18]   Kong H, Gao C and Yan D J. *Am. Chem. Soc.,* 2004,126, 412.

[19] Holzinger M, Vostrowsky O, Hirsch A, Hennrich F, Kappes M, Weiss R and Jellen F Angew. *Chem. Int.* Edn 2001,40, 4002.

[20] Holzinger M, Abraham J, Whelan P, Graupner R, Ley L, Hennrich F, Kappes M and Hirsch A J. *Am. Chem. Soc.,* 2003, 125, 8566.

[21] Yao Z, Braidy N, Botton G A and Adronov A 2003 *J. Am.Chem. Soc.* 125, 16015.

[22] Ying Y, Saini R K, Liang F, Sadana A K and Billups W E *Org. Lett.* 2003, 5, 1471.

[23] Alvaro M, Atienzar P, de la Cruz P, Delgado J L, Garcia H and Langa F 2004 *J. Phys. Chem.* B 108 12691.

[24] M R. Nyden, S I. Stoliarov, *Polymer.* 2008,49, 635-641.

[25] Rakhimkulov AD, Lomakin SM, Alexeyeva OV, Dubnikova IL, Schegolikhin AN, Zaikov GE in *Proceedings of IBCP International Conference* (2006) 56.

[26] Krusic J, Wasserman E, Keizer PN, Morton JR, Preston KF. *Science.* 1991, 254, 1183.

[27] Opfermann J. *J. Thermal Anal Cal.* 2000, 60, 3, 641.

[28] Friedman H.L., *J. Polym. Sci.,* 1965, C6, 175.

[29] Opfermann J., Kaisersberger E. *Thermochim. Acta.* 1992, 11, 1, 167.

[30] Lomakin SM, Dubnikova IL, Berezina SM, Zaikov GE, *Polymer Science.* 2006, 48, Ser. A, 1, 72.

[31] Gilman GW, Jackson CL, Morgan AB, Harris RH, Manias E, Giannelis EP, Wuthenow M, Hilton D, Phillips S, *Chem. Mater.* 2000, 12, 1866.

[32] Kashiwagi T, Harris RH Jr., Zhang X, Briber RM, Cipriano BH, Raghavan SR, Awad WH, Shields J R. *Polymer.* 2004, 45, 881.

[33] Kim P, Shi L, Majumdar A, McEuen PL. *Phys. Rev. Lett.* 2001, 87, 215502.

[34] Yi, W.; Lu, L; Zhang, D.L.; Pan, Z.W.; Xie, S.S. *Phys. Rev. B* 1999, 59, R9015–8.

[35] Echevskii GV, Kalinina NG, Anufrienko VF, Poluboyarov VA. *React. Kinet. Catal. Lett.* 1987, 33, 8, 305.

[36] Zhu. J, Uhl F, Morgan AB, Wilkie CA, *Chem. Mater.* 2001, 13,4649–54.

INDEX